Praise for
Science Was Wrong

"Stanton Friedman and Kathleen Marden, nuclear physicist and educator, have collaborated to show how often evolving scientific dogmas have been wrong and data that contradicted those dogmas was deliberately ignored. Science has been wrong about Jupiter, smallpox, medicines, food contaminations—and is wrong in its dismissal of the UFO phenomena that the authors show in Chapter 13 is a government-controlled policy of denial to make lies official truths."

—Linda Moulton Howe, Emmy Award–winning
TV producer and investigative journalist

"An excellent reference for errors in conventional science fields as diverse as astronomy, medicine, environmental science, psychic phenomena and UFOs."

—Dr. Bruce Maccabee, retired Navy physicist

"It's another hit for Friedman and Marden! The duo focuses their attention on outrageous scientific proclamations of the last 150 years, while highlighting potential knowledge breakthroughs Science refuses to consider and shunts aside—to the collective detriment of humanity."

—Rob Swiatek, physicist, UFO researcher

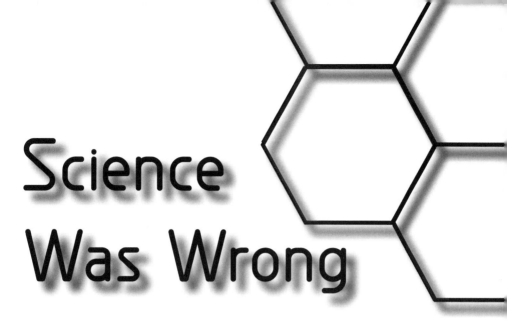

Science
Was Wrong

Startling Truths About Cures, Theories, and Inventions "They" Declared Impossible

By Stanton T. Friedman, MSc, and Kathleen Marden

A division of The Career Press, Inc.
Pompton Plains, NJ

SCIENCE WAS WRONG
EDITED AND TYPESET BY DIANA GHAZZAWI
Cover design by Dutton & Sherman Design
Printed in the U.S.A. by Courier

To order this title, please call toll-free 1-800-CAREER-1 (NJ and Canada: 201-848-0310) to order using VISA or MasterCard, or for further information on books from Career Press.

The Career Press, Inc., 220 West Parkway, Unit 12
Pompton Plains, NJ 07444
www.careerpress.com
www.newpagebooks.com

Library of Congress Cataloging-in-Publication Data
Friedman, Stanton T.
  Science was wrong : startling truths about cures, theories, and inventions "they" declared impossible / by Stanton T. Friedman and Kathleen Marden.
       p. cm.
  Includes bibliographical references and index.
  ISBN 978-1-60163-102-2
  ISBN 978-1-60163431-4 (ebooks)
    1. Science—Miscellanea. 2. Discoveries in science. I. Marden, Kathleen. II. Title.

Q173.F858 2010
500--dc22

22010003634

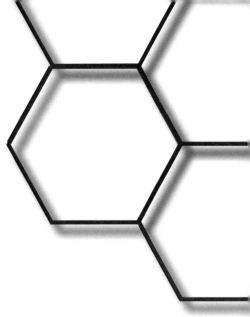

This book is dedicated to our loving spouses, Marilyn and Charles.

We wish to acknowledge the brilliant people who have voiced misguided proclamations of impossibility. Without them, this book would not have been possible.

Special thanks to Catherine Marden for her proofreading assistance and suggestions throughout the writing process and to Rob Belyea for preparing our photos for publication.

To our literary agent, John White, for successfully pursuing a publisher for our books.

To our editors at New Page Books, Kirsten Dalley and Diana Ghazzawi, for their expert assistance in this book's production.

And finally, to the individuals, too numerous to mention, who contributed to the successful completion of *Science Was Wrong* by providing expert opinions, releases, comments, and conference space. Your assistance is very much appreciated.

Contents

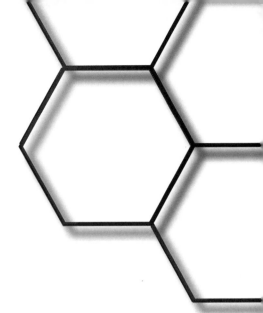

Politics

Frontiers of Science

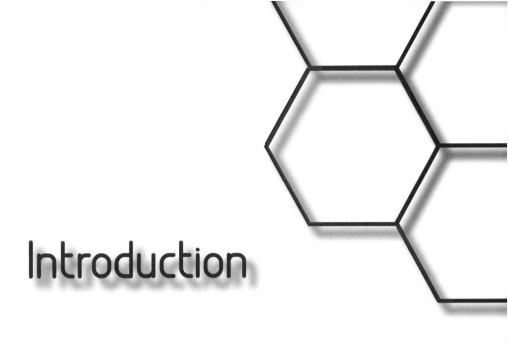

Introduction

The history of science and technology is rife with authoritative claims by reputable scientists that have impeded progress. Many of the claims seem humorous with the benefits of hindsight. There have also been a number of compilations of silly claims of impossibility made by smart people. But no book, until now, goes into the serious implications for society of blindly rejecting new scientific discoveries simply because they fail to conform to the preconceived paradigms of the scientific establishment. This book critically examines historical proclamations of "impossibility" in the areas of aerospace, technology, medicine, and politics that were later refuted, and investigates the frontiers of science.

The history of aerospace technology is loaded with well-connected scientists who resisted change. There were prominent "experts" who thought flight was not to be. Had they been supportive of progress in this field, there could have been real benefits to humanity. Could World War II have ended sooner if jet engines had been implemented earlier in England, where a patent had been granted in 1930? How many lives would have been saved if space travel had been pursued sooner, providing better advance information about natural disasters, such as tornadoes and hurricanes, and early warning of attacks from enemies?

Communication techniques didn't start changing until less than 200 years ago. The telegraph, telephone, television, Internet, and cell phone were all targets for the impossibilists. Sometimes it required true persistence to overcome the inertia of these traditionalists. Why would one want to sit in front of a box watching pictures? Of what use is a telephone if there is nobody to call? Their lack of imagination and foresight is obvious. Did they not think of the benefits of rapid, long-distance communication in medical situations or when ships at sea run into serious problems? Can we allow societal regulations to be established in response to pressures from people with a vested interest in continuing the status quo?

Germ theory was first advanced in ancient Sanskrit texts and later proposed in 36 BC. Despite observational and experimental data, it was not widely accepted until late in the 19th century, when Louis Pasteur discovered that microorganisms, not miasmas (the poisonous atmosphere arising from swamps and putrid matter), cause disease. Unfortunately, his discovery was not immediately accepted. Many influential opponents from the scientific establishment clung to their archaic beliefs and were slow to acknowledge that his germ theory of disease was valid. Attempts by his predecessors to impart evidence of the transmission of microscopic organisms as a cause of disease were largely unsuccessful. As a result, careers were ruined, and many people died because of a failure to implement new treatments and new understanding of various diseases. Even in our day, experts have often dismissed the dangers of new treatments long after the data was available. How many contracted HIV/AIDS because of the failure of governments to take appropriate measures? It is clear that innovative scientists have a long history of facing harsh rebukes by the medical establishment.

Worse, highly regarded but politically influenced scientists have promoted ideas that have led to human suffering. For example, social Darwinism fueled the Eugenics Movements in America and Germany and led to sterilization and extermination programs. The dark underbelly of corruption has reared its head in environmental science, causing thousands to die or be maimed due to methylmercury poisoning. Additionally, we are faced with environmental concerns about global warming. Should hundreds of billions of dollars be spent to attack evil carbon dioxide, or is this another example of vested interests triumphing over the real needs of society?

This book explores the frontiers of science, such as psychic phenomena and extraterrestrial visitation. A small group of vocal arch-skeptics claims a hundred years of research has failed to produce convincing evidence for psi phenomena. Parapsychologists disagree, arguing that hundreds of scientific studies have produced evidence that some psi phenomena are real. Each group accuses the other of confirmation bias. But what is the truth? Do parapsychologists selectively report evidence that supports psi phenomena? Or do arch-skeptics automatically reject statistically significant experimental replications?

For more than 60 years, strong attacks have been made on all aspects of the UFO question. Impossibilists have claimed that travel from other stars is impossible, there is no evidence for flying saucer reality, aliens could not possibly look humanoid, governments could not cover up the truth, aliens wouldn't behave the way they are supposedly observed to behave, eyewitness testimony is not part of the scientific method, there is no reason to abduct Earthlings, and so on. These claims are not derived from a serious review of the evidence, but rather are created by armchair theorizing without relation to the vast amount of evidence available for those who seek it out. False attempts have been made to show that Occam's razor supposedly rules out saucer reality, and that scientists in general—and astronomers in particular—do not observe UFOs. It turns out that the primary attacks on UFO reality by supposed professionals can generally be described as pseudoscientific. The attackers have inordinate confidence in themselves plus an almost religious faith in their feelings, intuition, and hunches. They simply don't need to investigate because they know the answers. That is not science, but pseudoscience.

During the past 200 years, there has been an enormous explosion of new discoveries and inventions in aviation, space, communications, medicine, and warfare. The claims of individuals who declared something was impossible help us to recognize how many lives have been lost and the benefits delayed by the impossibilists. All too often, pronouncements of impossibility were made from positions of ignorance rather than knowledge, but were listened to because of the exalted status of the speaker, rather than the evidence, if any, put forth in support of a position. Throughout history it has been difficult, if not impossible, to promote the acceptance of new discoveries. The reasons are multifaceted, but often involve arrogance and mountainous egos, politics and greed, and resistance to change. One can only wonder how many groundbreaking discoveries have been suppressed. There are many lessons to be learned by studying the various situations described in *Science Was Wrong*.

AEROSPACE

Between the time of the Wright brothers' first flight in 1903 and the launch of the first satellite (Sputnik) in 1957, there was an enormous growth in commercial, civil, and private aviation. Creative people were willing to disregard ridicule and the rants of the impossibilists. No longer did one hear that if God had intended for man to fly, he would have given him wings. Nevertheless, the impossibilists couldn't imagine the potential benefits of aviation and space exploration and their utilization.

It has always been clear that there would be a need for the expenditure of large amounts of money to develop the required infrastructure of a space program. After all, there were no landing facilities, refueling places, runways, weather information, communication systems, or regulations, but it didn't take long for governments to realize that there were strong military benefits—airplanes could be used for reconnaissance, defense, and bombing. The atomic bomb, developed in secrecy, was done with the objective of making a device that could be dropped from an airplane. The cost was an astronomical, for the day: two *billion* dollars. Of daily importance, the use of airplanes for mail delivery provided profit and an improvement on an existing national mail system. It also helped increase the size of planes and the frequency of flights.

With regard to space, many with limited imaginations couldn't foresee the benefits, such as weather forecasting, spying on one's enemies, and communications systems for accepting and receiving signals from the ground and redirecting them to other locations. New cameras, radio transmitters, and other devices had to be developed and miniaturized.

For whatever reason, astronomers were in the forefront of the impossibilist movements against both aviation and space travel. Perhaps it is because astronomers are normally not involved with, and could not justify, the expenditures of the huge sums of money that have been required for the development of high-performance aircraft, ICBM rockets for the delivery of nuclear warheads, very sophisticated cameras, and data transmission. Some of the spy satellites launched for the National Reconnaissance Office cost half a billion dollars each, which is more than the total budget of a host of astronomers. Yet, strangely, astronomy has been one of the fields benefiting most from new systems such as the Hubble Space Telescope, sophisticated devices landed on Mars and the moon, and such very special satellites as the Pioneer, Viking, Voyager, and Cassini spacecraft.

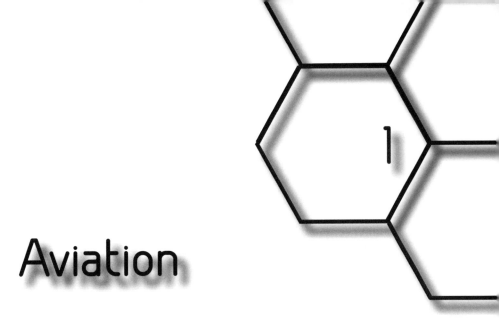

Aviation

Man has dreamed of flying for as long as he has observed birds soaring high. A number of brave (or foolish) people attached wings to their shoulders and tried soaring off hilltops...not very successfully. The ancient Greek legend of Icarus and Daedalus may have fueled some of these dreams. In the 13th century, Marco Polo reported that, while in China, he had observed large kites being used to "fly" people. During the Renaissance, Leonardo da Vinci sketched aerial devices, including a helicopter, though he was more focused on conceptualization than manufacturing or testing. The first manned flight actually used an unpowered hot-air balloon built by brothers Joseph and Jacques Montgolfier, paper mill owners, on October 15, 1783, in Annonay, France, and carried two passengers. Three months later, they carried seven passengers to a height of 3,000 feet. Unpowered gliders were being flown during the 19th century. They had no engines, but did have wings. Considerable aeronautics research was being done at the time, including by such pioneers as Sir George Cayley, who conducted many studies of aeronautics. His published results had a significant impact on the Wright brothers many years later.

There were, of course, prominent "experts" who thought flight was not to be. In 1896, the great Lord Kelvin, president of the British Royal Society, was offered a membership in what is now the Royal Aeronautical Society of Great Britain. His response was, "I have not the smallest molecule of faith in aerial navigation other than ballooning or of expectation of good results from any of the trials we hear of."[1] H.G. Wells, author of *War of the Worlds*, was almost as conservative, claiming in 1902 that "long before the year 2000 and very probably before 1950 a successful aeroplane will have soared and come home safe and sound."[2] It took less than two years.

Finally, on December 17, 1903, in Kitty Hawk, North Carolina, the Wright brothers (Orville and Wilbur—they took turns) made the first controlled sustained flights of a manned, powered vehicle. Perhaps not surprisingly, just two months before the Wright brothers' big day, one of the top astronomers of the 19th century, Dr. Simon Newcomb, had written, "The demonstration that no possible combination of known substances, known forms of machinery, and known forms of force can be united in a practical machine by which man shall fly long distances through the air, seems to this writer as complete as it is possible for the demonstration of any physical fact to be."[3] This seems to ignore the possibility that there would be new forms of machinery or new forces unknown to Dr. Newcomb, who was certainly not an expert on flying machine design.

The impossibilists seemed not to understand that generally technological progress comes from doing things differently in an unpredictable way, and that the future is rarely an extrapolation of the past. There was also a tendency to judge the newest-model device, on which research had just begun, against a very well-developed older technology. For example, newly invented cars were compared to well-developed trains. This was hardly a fair comparison. Newcomb was not only not involved in engineering work, but was also not acquainted with the work that had been done by others. His approach was very superficial, noting, for example, that the weight of a craft goes up more rapidly with increasing size than does the surface area. It turns out that the devil was very much in the details.

In contrast with Newcomb, Samuel Pierpont Langley was an engineer and architect with a strong interest in aviation. He had become secretary of the Smithsonian Institution in Washington, D.C., in 1887 and performed many studies of flight, including flying a number of unmanned, powered models, and measuring forces on flat and curved surfaces. In 1896, he had actually flown an unmanned, steam-powered, 14-foot model for 3,000 feet with surprising stability, even touching down gently when it ran out of fuel. Another model flew 4,200 feet. The War Department, in 1898, involved in the Spanish American War, gave him a grant for $50,000 (a considerable sum for those days) to try to achieve manned flight. He then built a full-sized, manned machine that had a 53-horsepower gasoline engine. He made two attempts using a ramp and catapult on the Potomac River to get it up to speed. Neither was successful, but at least the pilot wasn't killed when the planes crashed. There was much public ridicule after the second failure. The press referred to the debacle as "Langley's Folly." The Wright brothers' first flight was only nine days later, but done in obscurity rather than in the glare of publicity.

Of course, Newcomb wasn't the only naysayer. The great inventor Thomas Edison, who eventually held more than 1,000 patents, on November 17, 1895, was quoted as saying, "It is apparent to me that the possibilities of the aeroplane, which two or three years ago was thought to hold the solution to the flying machine problem, have been exhausted, and that we must turn elsewhere."[4] Strangely, Edison had been the target of impossibilists regarding the phonograph (it was claimed he must be a ventriloquist) and even the light bulb.

The Wright brothers were very skilled mechanics, though neither was a high school graduate, and designed and manufactured bicycles near Dayton, Ohio. They became interested in flight in the 1890s and gathered as much published information as they could from the Smithsonian Institution, many individuals, and a number of books that had already been published. They built a number of unmanned glider models but found that the measurements they made of important parameters, such as lift and drag, did not match the values predicted using the best available calculations. They then built their own small wind tunnel and made hundreds of tests until their calculations matched the measurements. They were well aware of the difficulties, such as finding a light-weight engine to drive a propeller (some had tried using heavy steam engines), selecting the right curvature of the wings, finding a means for turning the airplane, and landing safely. They had applied for a patent in March 1903, but didn't actually make flights in public until several years later. A total of five flights were made on December 17, 1903, the longest being 852 feet. Of course there were no TV cameras present at their first flight, and only three newspapers in the United States carried the news item. That very first aeroplane weighed all of 700 pounds, had an engine producing 12 horsepower, was a bi-plane (two wings), and with the first flight covered a distance of only 112 feet for 12 seconds. The entire flight would fit inside a 747 aircraft. It was a beginning, but hardly a harbinger of the progress soon to be made.

The brothers were hoping to get significant funding to improve their design. They applied for a government grant and were turned down twice. They finally got a contract for $25,000 for the purchase of a plane by 1908. They also received a bonus of $5,000 because the plane achieved a velocity of 47 mph as opposed to the required value of 40 mph. After 1905, they greatly reduced their flying and development, but brought a number of legal actions to protect their patent. They tried to make sales in Europe.

Their legal actions didn't stop Glenn Curtiss from building a large number of his own aircraft. He had been joined by Alexander Graham Bell, inventor of the telephone, in a new company. As a matter of fact, by 1916, Wright planes would constitute only 9 percent of the American Army aircraft, while Curtiss provided 69 percent. The Wright Brothers seemed to have become complacent with their success. Even Wilbur Wright wrote in October 1906, "We do not believe there is one chance in 100 that anyone will have a machine of the least practical useful-ness within five years."[5] Much sooner than that, the Europeans had overcome the early U.S. lead in aviation. A primary reason was the impending war and the recognition that in Europe, with a much larger population base and cities much closer to each other than in the United States, planes could be very useful. It is shocking to realize that in 1914, at the beginning of the First World War, Russia, Germany, France, and Britain had between them 751 aircraft. The United States had only 23.

Looking back from the vantage point of the 21st century, it is clear that military objectives and the willingness to spend huge sums of money were important in the development of many areas of technology including not only aviation, but space, submarines, communications, navigation, radar, and the miniaturization of electronic devices. Often these developments took place in spite of the lack of vision of many prominent politicians, scientists, and even military people themselves.

In the 1900s there was strong sentiment in certain circles against any optimistic thinking about the future of aviation, making it difficult to obtain funding. For example, on January 1, 1900, an engineer, Worby Beaumont, in response to a question from a journalist who had asked if man would fly in the next century, said, "The present generation will not fly and no practical engineer would devote himself to the problem now."[6] Rear Admiral George Melville, Engineer-in-Chief of the United States Navy, was quoted as saying in the *North American Review* of December 1901: "There is no basis for the ardent hopes and positive statements made as to the safe and successful use of the dirigible balloon or flying machine, or both, for commercial transportation or as weapons of war."[7] Dirigibles had already started flying in 1900. Some people could envision aircraft that would be much larger, carry more payload, and fly higher and faster than was possible then. Others couldn't. Even Octave Chanute, an aviation pioneer, stated in 1904 that "Airplanes will be used in sport, but they are not to be thought of as commercial carriers. To say nothing of the danger, the sizes must remain small and the passengers few, because the weight for the same design will increase as the cube of the dimensions, while the supporting surfaces will only increase as the square."[8] Chanute was very much sold on aviation. He published freely, and consulted with the Wright brothers. He had been a very successful engineer building bridges, railroads, and other structures.

It is interesting that though American engineering was in the lead early on, as World War I approached, Europeans led the way. When American pilots joined up, they were flying mostly European planes. Reconnaissance and artillery target–spotting were, of course, one benefit, but it was found that as engines got more powerful and payloads greater, other uses were feasible. Guns could be mounted on airplanes, and it was possible to communicate between the air and the ground and plane to plane. Bombs could be dropped, as well. However, one estimate of the situation by Charles Stewart Rolls, automobile and aviation pioneer and cofounder of Rolls-Royce (still a major manufacturer of aviation engines), in 1908 stated, "I do not think that a flight across the Atlantic will be made in our time, and in our time I include the youngest readers."[9] That feat was achieved only 11 years later in June 1919 by the Vickers Vimy IV bomber, flown by John Alcock and Arthur Whitten Brown from Newfoundland to Ireland.

By the end of World War I, the United States had produced a number of airplanes, including large seaplanes, and there was serious discussion of crossing the ocean. In 1915, the United States finally set up a research and development organization, the National Advisory Committee on Aeronautics. Among their interests were the military activities that could be conducted. For example, planes could be used to shoot down enemy aircraft, negating their reconnaissance efforts. Some people were envisioning transatlantic flights by seaplanes that would land on the water, be refueled by a waiting tanker, and then take off again.

American General Billy Mitchell, who was very active in military aviation in Europe during the First World War, after the war was espousing the notion that airplanes would change the nature of warfare. He also made the extraordinary claim to some that airplanes would be able to sink ships. Newton D. Baker, U.S. Secretary of War, stated in 1921, "The idea is so damned nonsensical and impossible, that I am willing to stand on the bridge of a battleship while that nitwit tries to hit it from the air."[10] He would have died of drowning when the German battleship *Ostfriesland* was sunk by bombs dropped from the air in July 1921. The Secretary of the Navy, Josephus Daniels, had joined the chorus with this sage comment: "Good God, this man Billy Mitchell should be writing dime novels."[11]

The same fatalistic attitude was still in place in 1941. The caption to the photograph of the battleship USS *Arizona* in the program of the Army-Navy Football game of November 29, 1941, stated, "It is significant that despite the claims of air enthusiasts, no battleship has yet been sunk by bombs." It was only eight days later that the USS *Arizona* was itself sunk by Japanese bombs dropped from the air at Pearl Harbor, Hawaii, killing 1,102 men. It seems obvious that a bomb dropped on a ship needn't completely destroy or vaporize the ship, but rather produce a hole or break a welded seam so that water can flow in, and the weight of the water replacing air would do the sinking.

Billy Mitchell was court-martialed in 1925 after accusing the Army and Navy of incompetence in various affairs. According to the trial, he was strong minded, spoke his mind, and was insubordinate. It turns out he was also essentially correct. When his son asked many years later that the conviction be overturned, he was turned down. Being right obviously wasn't enough.

Sir John Alcock and Sir Arthur Whitten Brown.
Courtesy of Arpingstone at Wikipedia Project.

The rapidly growing aviation role in the First World War led to continued aircraft development with planes that could fly faster, higher, farther, and with more payload than ever before. This was despite those people who had felt there was little future to aviation. Certainly one of the most significant events in aviation history was the widely publicized flight by American pilot Charles A. Lindbergh from New York to Paris on May 20–21, 1927. It was the first solo nonstop flight across the Atlantic. It lasted for 33.5 hours and covered 3,500 miles. Economic times were good (until the stock market crash of October 1929) and Lindbergh, only 25 at the time, received worldwide acclaim. The trip wasn't an easy one, as he had to fly as low as 10 feet above the ocean and as high as 10,000 feet, was in fog for a good while, had problems with wings icing up, and was navigating by the stars. Six other pilots had died in trying to win the $25,000 Orteig Prize for the first pilot to fly nonstop and solo between New York and Paris. There were 150,000 people at Le Bourget airport in France when he landed, though only 500 watched him leave New York. He was given many honors, such as the Congressional Medal of Honor. Passenger-carrying airlines progressed rapidly in the United States because of the enormous publicity concerning his flight. In 1926, they carried 5,782 passengers. In 1929, the total had soared to 173,405. The number of licensed aircraft quadrupled. The number of applications for pilot's licenses tripled. Air mail service also grew rapidly.

General Billy Mitchell. Courtesy of Arpingstone at Wikipedia Project.

It is interesting that Lindbergh was allowed to review German aircraft and other military developments in the 1930s. He warned that the Germans had been making considerable progress in aviation. As a leader of the non-intervention American First Organization, he tried to keep the United States out of the European war. He changed his mind after Pearl Harbor. He was not allowed to reclaim his commission as a colonel, but served in an advisory capacity as a civilian though actually flying a number of military missions, estimated at 50.

One major goal in the 1930s, besides the development of all-metal construction, was the improvement in the power-to-weight ratio of engines. Englishman Frank Whittle, a pilot, received a patent on a jet engine as early as 1930. There was strong resistance on the part of the UK government research and development community, partly on the basis that gas turbines used in power plants and elsewhere were indeed quite heavy and were often over designed. They were not designed to be lightweight, as would be required for an aircraft engine. In addition, the higher the temperature at which they operate, the higher the efficiency and the lower the weight-per-unit power. But it seemed unlikely to some that engines could be made to operate at very high temperatures. A German inventor and physicist named Dr. Hans von Ohain had patented a jet engine in Germany in 1936, six years after Whittle's patent. Yet the German Heinkel HE178 airplane was the first jet-propelled aircraft to fly, using his jet engine, in 1939. The first English plane to use a jet engine wasn't flown until 1941, because of the lack of vision of English military "experts." Manufacture of the Whittle engine

Charles Lindbergh. Courtesy of the Library of Congress.

was moved to the United States, away from German bombardment, and General Electric began producing useful jet engines at the end of the war. If the development of the Whittle engine had been encouraged early on, rather than discouraged, many lives might have been saved and the war may have ended sooner. Jet fighter planes, had they been developed earlier, could have helped defend England against German bombers. It is of interest that, in addition to his many German contributions, Dr. Von Ohain came to Wright-Patterson Air Force Base in Dayton, Ohio, in 1947, and had a long and very distinguished career, receiving many honors for his major contributions to aeronautics.

Early on there had been reluctance to pursue the aircraft as a weapon of war. In 1908, Harvard astronomer William H. Pickering stated, "A popular fallacy is to suppose that flying machines could be used to drop dynamite on an enemy in time of war."[12] Furthermore, the well-respected *Scientific American* on July 16, 1910, claimed, "To affirm that the aeroplane is going to revolutionize naval warfare of the future is to be guilty of the wildest exaggeration."[13] Marechal Foch, a famous French Army general, said in 1911, "Airplanes are interesting toys, but of no military value."[14] Even in 1921, the U.S. Secretary of War, John Wingate Weeks, said, "[Army General] Pershing won the world war without even looking into an airplane let alone going up in one. If they had been of such importance, he'd have tried at least one ride. We'll stick to the army on the ground and the battleships at sea."[15]

In 1932, an article in *The American Mercury* claimed, "Air forces by themselves will never do to great cities what Rome did to Carthage or what the Assyrians did to Jerusalem."[16] As late as 1937, a U.S. Marine Corps colonel, John W. Thomason, said, "It is not possible to concentrate enough military planes with military loads over a modern city to destroy that city."[17] To say that the potential for destruction of cities using large numbers of bomber planes was vastly underestimated is not an exaggeration. Many cities in Germany and Japan were devastated by massive air raids involving hundreds of bombers. During World War I, most battles involved armies facing each other across trenches. During World War II, the destruction of civilian targets was the big objective. For example, on March 9–10, 1945, 279 B-29s dropped 1,700 tons of bombs on Tokyo, killing an estimated 88,000 people, injuring 41,000, and destroying 16 square miles. In Germany, 500,000 people were killed and 7 million homes destroyed in 161 cities. On November 14, 1940, 500 German bombers attacked Coventry, England, for 10 hours. The planes grew bigger and could carry more destructive bombs. Toward the end of the war, B-29s could carry 10-ton blockbusters. That is a lot of wallop in one bomb, though it can't compare with atomic or hydrogen bombs.

Of course, there had been resistance to the development of atomic bombs as well. Albert Einstein had stated in 1932, "There is not the slightest indication that nuclear energy will ever be obtainable. It would mean that the atom would have to be shattered at will."[18] Lord Ernest Rutherford, an outstanding English physicist, had claimed in 1933, "The energy produced by the atom is a very poor kind of thing. Anybody who expects a source of power from the transformation of these atoms is talking moonshine."[19] The power didn't come from transformation, but an entirely new process, nuclear fission. Neither scientist was aware that the neutron, which wasn't even discovered until 1932, and nuclear fission, first noted in 1938 by German scientists Otto Hahn and Fritz Strassmann, would lead to a controlled nuclear chain reaction in 1942 by Enrico Fermi, and, eventually, after the expenditure of $2 billion in 1942, to the development of the atomic bomb in great secrecy. A key discovery by Lise Meitner and Otto Hahn was that each fission produced about 2.5 more neutrons. This allows for a controlled, steady nuclear chain reaction, as in powerful nuclear reactors, and in a very rapid chain reaction, as in an atomic bomb. Yes, it was indeed possible to shatter the atom at will. Even Admiral William Leahy, advising President Truman, who had not been informed of the bomb until the death of President Roosevelt, stated in early 1945, "This is the biggest fool thing we have ever done. The bomb will never go off, and I speak as an expert in explosives."[20] The first nuclear test explosion, at the Trinity site at the White Sands Missile Range in New Mexico, on July 16, 1945, provided energy equivalent to that in 12,000 tons of TNT. The bombs dropped by air on Hiroshima and Nagasaki in August 1945 released a similar amount of energy. It only took seven more years before the first hydrogen bomb (powered by nuclear fusion, using isotopes of hydrogen) was exploded in the Pacific. It released energy equivalent to 10 million tons of TNT and had a fireball three miles wide. Soon such weapons could be carried by large bombers. The ability to pack so much punch in a small package had, of course, made it obvious that a long-range missile could be used to rain destruction on a distant enemy.

> Einstein, Rutherford, Strassmann, and Hahn all won Nobel Prizes in Physics, as did Fermi, who was heavily involved in nuclear weapons development.

It was certainly expected by certain engineers and scientists that atomic energy could also be used to produce electrical power, nuclear-powered submarines, and nuclear-powered aircraft carriers. Today, submarines can travel under water for years, and aircraft carriers can operate without refueling for 18 years. Both are excellent examples of progress coming from doing things differently.

Not only did the impossibilists slow down our progress, but they also, for whatever reasons, consistently underestimated Soviet capabilities. General Leslie R. Groves, who was in charge of the Manhattan Project, the Army effort to develop atomic bombs, stated "optimistically" on June 19, 1948, in the *Saturday Evening Post*, "It will take Russia at least until 1955 to produce successful atomic bombs in quantity. I say this because Russia simply does not have enough precision industry, technical skill, or scientific numerical strength to come even close to duplicating the magnificent achievement of the American industrialist, skilled labor, engineers and scientists who made the Manhattan Project a success." It turns out their first atomic bomb was exploded in August 1949, to the great shock of the American political and military establishment. Furthermore, the U.S. military thought they were still safe for quite some time, because the Russians supposedly didn't have planes to deliver the atomic bombs. However, they had quite successfully copied an American B-29 bomber that had been left in Russia during the war and built atomic bombs and the aircraft for delivering them much sooner than expected.

In the late 1950s, Convair developed the first supersonic bomber the B-58 Hustler. It was a swept delta-wing plane that could achieve twice the speed of sound in level flight. Unfortunately, many of the 116 built wound up in storage, because they couldn't compete with the much slower B-52s and B-47s in terms of payload and range.

Though it was known that fighter planes in dives from high altitude could get to very high speeds, there was a price to pay in terms of air flow and serious vibration developing. The notion of a sonic barrier as the plane approached the speed of sound (700 mph at sea level) became a problem. The great majority of "experts" believed that the sonic barrier could not be breached. Of course they were thinking of older, conventional planes affixed with propellers. To help solve the flow problem, Richard Whitcomb devised the "coke bottle" shape aid during the transition between subsonic and supersonic flight speeds. Test pilot Chuck Yeager broke the sound barrier on October 14, 1947, in the Bell XS-1 Rocket plane. It was carried aloft beneath a B-29 bomber, and, at an altitude of 35,000 feet, was dropped from the plane. It was aimed upward at an angle and gradually increased its speed until it reached Mach 1.06 at 43,000 feet, and eventually glided to a landing on the desert runway. The plane only weighed 5,000 pounds empty, but carried 8,000 pounds of propellant—enough for less than three minutes of flight. The test was done in secret, with the news that the sonic barrier had been broken not being released until later. It didn't take too long until more and more jet fighters were built, tested, and flown as the Cold War took over military thinking.

The United States did not build the first civilian jet-powered transport either. That honor went to England, which developed the Comet. It operated for some time before it had several serious accidents, apparently caused by unanticipated metal fatigue resulting from expansion and contraction when flying at high altitude and then at low altitude.

There was also a considerable effort devoted to the development of a nuclear-powered airplane during the 1950s. Such a plane could be flown for huge distances and lengths of time without refueling, if the reactor–engine combination could be made small and light enough. There would be no midair refueling and no restriction on range. There were, of course, serious concerns about safety in the event of an accident involving a powerful nuclear reactor. More than $1 billion were spent and significant progress was made in the development of high-performance reactors, radiation shielding, high-temperature materials, as were required for the reactor and turbines and shielding and jet engines were operated on the ground on nuclear power. But no nuclear powered airplane was flown. The program was finally cancelled in 1961. It seems surprising to many that the General Electric Aircraft Nuclear Propulsion Department in Evendale, Ohio, spent $100 million in 1958 alone. It employed 3,500 people, of whom 1,100 were engineers and scientists, yet most people were not aware of the program. A nuclear ramjet was also developed and ground tested by Ling Temco Vought. It was operated on the ground, but not flown. It would have been carried aloft by a normal airplane, and then cranked up and would have been able to drop nuclear weapons on any place in the world.

Many other programs were conducted in great secrecy and at great cost. For example, the F-117 Stealth aircraft was developed by Lockheed at a cost of $10 billion during a 10-year period before the very existence of the program was revealed. The B-2 bomber was also developed in secret and at great cost. Many rumors have circulated about anti-gravity systems and tapping the energy of the vacuum. Impossibilists have not had access to the data—if there is any. In addition, there was considerable secret effort related to the possible use of magneto-aerodynamic systems that involved ionizing the air around a craft and using electric and magnetic fields to control lift, drag, heating, sonic-boom production, and radar profile. A literature search conducted in 1970 of technical reports developed under government contracts found 900 reports using the key word "magneto-aerodynamics," of which 90 percent were classified. Some undoubtedly related to the problem of rocket nose cones entering the atmosphere at high speed and influencing lift and drag. Strangely, an electromagnetic submarine was actually built and tested by Dr. Stuart Way of Westinghouse Research Labs and a number of students at the University of California, Santa Barbara in the mid 1960s, and by a Japanese industrial group somewhat later. (Seawater is an electrically conducting fluid, so there is an analogy to an airborne system in which the air is made electrically conducting. If work was done in these areas more recently, it hasn't been released.)

It should not be surprising, based on the many false pronouncements by well-educated impossibilists about aviation, that a similar attitude has often been expressed toward all aspects of space travel. These are the subject of Chapter 2.

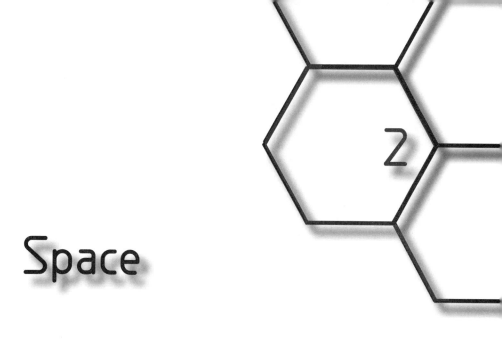

Space

Throughout the years, there have been many claims of impossibility for space travel, both manned and unmanned, within the solar system and beyond to the stars. This is not surprising in view of the resistance to the possibility of flying powered aircraft just within the atmosphere, as described in Chapter 1. Everybody was well aware that birds and insects could fly, but machines were another matter. Science fiction writers such as Jules Verne (who wrote *From the Earth to the Moon* in 1865) and H.G. Wells (who wrote *War of the Worlds* in 1898) provided stimulation to many about space travel, but they were the exceptions.

The life of American rocket pioneer Dr. Robert Hutchings Goddard provides solid examples of the impact of ignorant criticism from supposedly well-educated impossibilists. He had been inspired as a youth by both Verne's and Wells's books. As early as 1902, while still a student, he had submitted an article, "The Navigation of Space," to *Popular Science News*. He received a doctorate in physics at Clark University in Worcester, Massachusetts, in 1911, and, from then on, did rocketry research, as well as taught physics. As early as 1914, he had patented various rocket component designs.

Dr. Robert H. Goddard.
Courtesy of NASA.

Goddard worked for the Army in 1917 to try to develop devices that might help win the war in Europe. He actually succeeded in designing several systems that could have been used in the trenches, and had demonstrated solid-fueled rockets at the Aberdeen Proving Ground on November 7, 1918. This was four days before the war ended, and the Army never ordered any. He continued on with his research. In 1919, he published a very scientific paper, "A Method of Reaching Extreme Altitudes," in the *Smithsonian Miscellaneous Collections* (v. 71, no. 2, 1919). It covered much of the research he had done prior to that time and included the possibility of space flight. Few understood his work and he was dubbed "Moon Man." The *New York Times*, in a January 13, 1920 article, ridiculed him with this really nasty and totally incorrect comment: "That Professor Goddard, with his 'chair' in Clark College and the countenancing of the Smithsonian Institution, does not know the relation of action to reaction, and of the need to have something better than a vacuum against which to react—to say that would be absurd. Of course, he only seems to lack the knowledge ladled out daily in high schools."[1] Of course, the *Times* was totally wrong. Rockets work by their exhaust reacting against the rocket, not the surrounding air or vacuum. The newspaper took its own sweet time correcting its comments. Finally, on July 17, 1969, while *Apollo 11* was on the way to the moon, the *Times* provided this statement: "Further investigation and experimentation have confirmed the findings of Isaac Newton in the 17th century and it is now definitely established that a rocket can function in a vacuum as well as in an atmosphere. The *Times* regrets the error."[2]

Of course they didn't regret the damage done to the field of rocketry and the false attacks on Goddard. Goddard was badly hurt by these early completely unwarranted attacks and vowed to do his launches away from the glare of publicity. Much of that work was done on the Mescalero Ranch near Roswell, New Mexico, during the years between 1930 and 1941. He focused on liquid-fueled rockets and the improvement of pumps, valves, and other components. His liquid oxygen/ gasoline rocket, launched on March 16, 1926, was the first ever successful launch of a liquid-fueled rocket. This was only 13 months before Charles Lindbergh's solo transatlantic flight of May 1927, which gave an enormous impetus to aviation. Lindbergh later helped get support for Goddard from the Guggenheim Foundation.

Goddard fired his last of many rockets near Roswell in 1941. He provided loads of data to the military, but they simply weren't interested in long-range rockets. They sought his assistance in developing Jet Assisted Take Off (JATO) devices to help aircraft get off the ground in a hurry, and did use these. In very, very belated recognition of his research, in 1960, NASA named its spaceflight facility in Greenbelt, Maryland, The Goddard Space Flight Center, and paid his widow $1 million for the use of his more than 200 patents related to rocketry. He had died, largely unheralded, on August 10, 1945.

After World War II, it was learned that the German rocket pioneers under Werner von Braun had been well aware of Goddard's work, and used much of what he had published in their earlier development of the deadly V-2 rocket with which they had bombarded England.

In parallel, and independently, there was also the work of Konstantin Tsiolkovsky in the Soviet Union. In 1924, he heard about Goddard's 1919 paper and republished many of his more than 500 papers on space topics. Tsiolkovsky's "Exploration of Cosmic Space by Means of Reaction Devices" in 1903 showed that escape velocity of 5 miles per second could be reached by a multistage rocket fueled by liquid oxygen and liquid hydrogen. He received recognition during his lifetime and was buried in state.

Also in parallel, Hermann Oberth, a German rocket pioneer, was a leader in the German efforts to build rockets between 1930 and

1945. Werner von Braun was his pupil and later his boss when Oberth worked in the United States. He is often referred to as the "Father of Modern Rocketry." Perhaps not surprisingly, Oberth's doctoral dissertation on rocket science was actually rejected in 1922 by an impossibilist as "Utopian."

These three worked independently, but all published and made significant leaps in rocketry, making it difficult to justify some of the silly attacks by some "scientists." For example, Professor Alexander Bickerton, an astronomer from New Zealand, presented a paper at the 1926 meeting of the British Association for the Advancement of Science in which he "scientifically" showed that it would be impossible to give any object sufficient energy to go into orbit around the Earth. He was obviously completely unfamiliar with the

Dr. Werner Von Braun. Courtesy of NASA.

work of such real rocket scientists as Goddard, Tsiolkovsky, and Oberth. He considered the amount of kinetic energy that would be possessed by an object moving at escape velocity and noted that, using the best explosive, nitroglycerine, he could release only 1/10 as much energy per pound as would be required. He obviously didn't understand that it was the payload that was to be placed in orbit, not the propellant, and that, using his technique, it would take the energy in 10 pounds of explosive to place one pound of payload in orbit. Furthermore, he obviously wasn't aware that a rocket using liquid oxygen and liquid hydrogen could release much more energy per pound, and was much more manageable than one using an explosive.

Bickerton was but one of many astronomers claiming impossibility and foolishness with the notion of rockets for space flight. Another excellent example of impossibilities was the paper "Rocket Flight to the

Moon," published in the January 1941 issue of *Philosophical Magazine* by Dr. John W. Campbell, a Canadian astronomer. Upset by the same science fiction stories that had inspired Goddard and Oberth, Campbell set out to show how impossible it would be to send a man to the moon and back. His conclusion, after pages of equations and computations, was that the required initial launch weight of such a chemical moon rocket would be a million million tons. As is usual in such cases, he was apparently totally unaware of the work of Goddard, Oberth, and Tsiolkovsky. When, in 1969, the United States sent three men to the moon and back in a Saturn 5 chemical rocket, it weighed all of 3,000 tons at launch. As it happened, Campbell seems to have made every possible wrong assumption, clearly indicating his total ignorance of a huge body of work. For example, he assumed a single-stage rocket would be used, though all deep-space rockets have been multistage. The idea is to throw away the big fuel container of the first stage (and then the second stage), so it need not be accelerated with the later stages. He assumed much too low an exhaust velocity for the propellant. He also assumed a limit of 1 G acceleration for the rocket. In actuality, astronauts routinely take 6 Gs and often as much as 16. The greater the acceleration, the quicker one gets to orbit and the less penalty one pays to the gravitational field of the Earth which is pulling the rocket back.

Of course, Campbell, in his ignorance, made no effort to take advantage of Mother Nature. By launching to the east from near the equator, he could have gained a "free" 1,000 miles per hour because of the Earth's rotation. Using a launch time selected to put the rocket at the right location to use the moon's gravitation field, he could have used that to add to his acceleration. Furthermore, in a major crazy assumption, he assumed that the only possible way to slow the rocket down as it approached the Earth at 25,000 miles an hour returning from the moon would be to fire a retrorocket. Unfortunately, it would, of course, have to have been launched from the Earth, slowed down at the moon before landing there, and then again accelerated away from the moon and decelerated back here. Real rocket scientists have recognized that there is another way: use the Earth's atmosphere to slow the landing vehicle down at no cost in propellant. The Apollo 13 mission demonstrated how important it was to get the reentry angle correct. Being clever was much more important than being powerful.

It is not at all clear why astronomers have been impossibilists about flight within the atmosphere and in space. Nothing in their training, education, and professional activities has anything to do with the design, fabrication, or operation of a rocket or space craft. They also certainly would not be expected to have any special insight into the motivations of explorers, visionaries, profit-seeking capitalists, and military-intelligence types. Human motivation is not the long suite of astronomers, nor are they accustomed to dealing with large-scale industrial activities involving billions of dollars and often done in secrecy. A large research program in terrestrial astronomy might involve dozens of people and a few million dollars, as opposed to the multi–billion dollar classified space-travel programs, each involving thousands of people. Spy satellites and communication and meteorological devices, for example, have proven to be of extraordinary importance, as have GPS systems—but not to astronomers.

The British Astronomer Royal Sir Richard van der Riet Wooley made many silly claims about space flight, including, when speaking to *Time* magazine for the January 16, 1956 issue, "It's utter bilge. I don't think anybody will ever put up enough money to do such a thing. What good would it do us? If we spent the same amount of money on preparing first-class astronomical equipment, we would learn much more about the universe...it is all rather rot." This was one year before *Sputnik*, and 13 years before the first manned landing on the moon. Isn't it ironic that astronomy has been so enriched by data obtained by such extraordinary space-based observatories as Hubble, Chandra, Fermi, Spitzer, Hipparchos, and many more?

Another example of an impossibilist involved a strange comment by Dr. Lee DeForrest, inventor of the vacuum tube, which made radio practical. He claimed on February 25, 1957, as reported in the *St. Louis Post Dispatch*, *New York Times*, and elsewhere, that the notion that man would go to the moon, land, pick up rocks, and bring them back to Earth for analysis was the wildest science fiction worthy of a Jules Verne: "I am bold enough to state that, no matter what the discoveries of the future, it will never happen." It took all of 12 years. It did not take new fundamental discoveries, but hard work by highly motivated engineers spending $20 billion and recognizing that technological progress comes from doing things differently. There is no reason to believe he knew anything about aeronautics and astronautics. Believe it or not, De Forrest had himself been the target of legal actions for trying to sell what "must have been phony stock" in a company actually claiming that radio signals could be sent across the ocean. The impossibilists had of course denied that possibility.

It is much more difficult to understand the more recent antagonism to deep space flight from such well-known younger scientists as Dr. Lawrence Maxwell Krauss, professor at Arizona State University's School of Earth and Space Exploration, and director of the A.S.U. Origin Initiative. An example of his ignorance about space flight is this comment from a *New York Times* piece he published on April 30, 2002: "We may not know how spacecraft of the future will be propelled, whether matter-antimatter drives will be built or even if time travel is possible. But we do know absolutely, how much on-board fuel will be needed to speed up a substantial spacecraft to near the speed of light—an enormous amount, probably enough to power all of human civilization at the present time for perhaps a decade." This claim is totally unsubstantiated and ridiculous. He certainly gives no clues in his books *The Physics of Star Trek* (1995) or *Beyond Star Trek: Physics From Alien Invasions to the End of Time* (1997) that he is any more accurate than Dr. Campbell was in his computations! His logic is akin to asking how long it takes to go from New York to San Francisco. The answer is reliant on whether one walks or goes by bicycle, car, train, or plane. The short answer is that everything depends entirely on the assumptions made.

It took Ferdinand Magellan's ship three years to go entirely around the Earth back in 1522. Jules Verne talked about going around the world in 80 days in a balloon in the late 1800s. Dick Rutan and Jeana Yeagher went around the world nonstop and unrefueled in their Voyager airplane in 216 hours in 1986. Steve Fossett did it in a hot air balloon in 67 hours in 2005. The space station and many other satellites take all of 90 minutes. The devil is certainly in the details. One might also have thought that the notion of going around the world underwater was absurd 70 years ago. The *Triton* nuclear-powered submarine did it in 1960.

In *Beyond Star Trek*, Krauss commented about possible advanced propulsion systems: "The most likely candidates in the short term are nuclear thermal rockets currently being studied by a group at NASA's

Lewis Research Center near where I live in Cleveland, Ohio." He seemed totally unaware of the many successful ground tests of nuclear thermal rockets in the 1960s in the KIWI, NERVA, Phoebus, and XE-1 programs conducted by Los Alamos, Westinghouse Astronuclear Lab, and Aerojet General. These tests were run at a nuclear test site west of Las Vegas. The most powerful system was the Los Alamos Phoebus 2B, less than 7 feet in diameter, which operated at a power level of 4,400 million watts, or twice the power of the Grand Coulee Dam. In theory, such systems are simple: use very cold liquid hydrogen to cool the reactor and then be exhausted at very high temperature to produce thrust.

On the next page of his book, Krauss spoke of nuclear electric propulsion systems: "However, if the political climate changes and the use of nuclear reactors in space becomes acceptable..." It seems he was completely unaware of the fact that about three dozen nuclear reactors had already been operated in space. All but one were launched by the Soviet Union. One even received enormous publicity when the Cosmos 954 space satellite, with a reactor on board, crashed in Northern Canada in 1978. Krauss also doesn't mention that many satellites have been powered by small radioisotope thermoelectric generators such as have been used on the *Pioneer, Voyager, Transit*, and *Cassini* spacecraft. They have no moving parts and last for ages.

Krauss admitted that nuclear fusion (which powers the stars and H-bombs) might be an interesting idea for propulsion, but wonders, "How do you ensure that all the energy goes out the back and doesn't melt the engine?" Again he was completely unaware of the many studies done in England (Project Daedalus) and the United States. These looked at the deuterium-helium-3 fusion reactions, which have the great advantage of not only producing much more energy per reaction than the deuterium-deuterium or hydrogen-hydrogen reactions noted by Krauss, but producing almost all their energy in the form of charged particles, rather than neutrons. The charged particles can be directed by magnetic and electric fields to exhaust out the back end of the rocket with about 10 million times as much energy per particle as can be obtained in a chemical rocket. An excellent paper, "Controlled Fusion Propulsion," was published in 1962 by John Luce and John Hilton at Aerojet General Nucleonics at the conclusion of a $9 million study contract for the U.S.

Air Force. Obviously Krauss had not done his homework, especially with regard to research performed outside academia in large programs.

A noted astronomer, Dr. Neil de Grasse Tyson, director of the Hayden Planetarium at New York's American Museum of Natural History, claimed on the Peter Jennings ABC documentary *Seeing Is Believing* of February 24, 2005, that star travel was impossible because our fastest spacecraft, *Voyager*, would take 70,000 years to get to the nearest star. Of course, it has no propulsion system and has been basically coasting since it left Earth. A balloon doesn't tell us much about high-speed aircraft any more than throwing a bottle into the ocean would tell us much about transatlantic trips by the *Queen Mary 2*.

Finally, it is useful to read the article by physicist Dr. Edward Mills Purcell in the book *Interstellar Communication: The Search for Extraterrestrial Life*. He looked at a relativistic trip to a star only 12 light-years away using matter–antimatter annihilation, and concluded "all this stuff about traveling around the universe in spacesuits—except for local exploration which I have not discussed—belongs back where it came from, on the cereal box."

Purcell had a very impressive background. He earned a PhD in physics from Harvard in 1938, worked at the MIT Radiation Lab in the early 1940s on electronics and communications research (undoubtedly classified), and won a Nobel Prize in physics. As a professor at Harvard in 1952, he worked with Dr. Ernest Bloch on new methods for nuclear magnetic precision measurements. This was a way to detect the extremely weak magnetism of the atomic nucleus. This work was the eventual basis for the development of magnetic resonance imaging (MRI). He and a graduate student, Harold I. Ewen, also were the first to detect electromagnetic waves from clouds of hydrogen in space at a frequency of 1,420 megacycles (wavelength of 21 cm). This has become the basis for most radio telescope searches for signals from extraterrestrial civilizations (such as SETI).

Purcell's paper "Radio-Astronomy and Communication Through Space" was first presented at Brookhaven National Laboratory in Long Island in 1960. The written version is one of 32 papers in *Interstellar Communication: The Search for Extraterrestrial Life*. It is basically an anthology, and has two papers by Frank Drake and many others by SETI specialists. As one might expect, there are no papers discussing the worldwide evidence concerning alien visitations to Earth such as are discussed in Chapters 13 and 14 of this book.

Only about five pages of Purcell's paper deal with interstellar travel. As might be expected, this portion is very simplistic, with little concern with the "how" details. He assumes matter–antimatter annihilation at 100-percent efficiency, a maximum velocity of only .98c (c is the speed of light), a target star 12 light-years away, acceleration at 1 G half way out and deceleration by 1 G the second half of the trip, and the repeat of the process in reverse to come home. The fraction of the total huge initial launch weight available for payload is very, very small, hence the "cereal box" comment. Purcell assumes the entire mass of the fuel would be launched from Earth, rather than refueling at the target star or some intermediate propellant refueling station. He ignores the best nuclear fusion reactions, such as deuterium-helium-3, which provides five times as much energy per reaction as the hydrogen–hydrogen reaction he assumes. The sun and all other stars produce their energy using nuclear fusion reactions. The proper reactions, as noted, provide charged particles that are born with 10 million times as much energy per particle as in a chemical rocket. These can all be directed out the rear of the rocket.

Of particular importance is the fact that at 1 G, it only takes a year to get close to the speed of light; many scientists have guessed it would take a decade or century at 1 G to reach the speed of light. Once one achieves a cruising speed of, say, .9999c, one would coast rather than keep accelerating. At that velocity, it only takes six months pilot time to go 37 light-years. The benefits of Einsteinian time slowing down get to be very much more important at speeds very close to c. For those who might wonder about such things, the new, huge, Large Hadron Collider particle accelerator in Switzerland collides beams of particles traveling at 99.9999 percent of the speed of light.

Dr. Freeman Dyson, who worked on the Orion deep space propulsion system, has written a paper, "Gravitational Machines," in which he notes the use of the gravitational field of one star to provide acceleration to an object sent from a nearby star. He claims that a binary pair of white dwarf stars could accelerate a spaceship by 10,000 Gs without any internal problems for the crew and craft.

A German scientist, Sebastian von Hoerner, also contributed a paper, "The General Limits of Space Travel," to _Interstellar Communication_. He has more equations than does Purcell, but, again, no details about specific flight profiles such as an engineer would normally review. He, too, comes up negative, though he was not as far off as Dr. Campbell. There is no suggestion in the 1960 book of near-future trips to the moon, of the landing of probes on Mars, and our traipsing out past Jupiter, Saturn, Uranus, and Neptune.

It is of some interest that Purcell had also been a member of the Science Advisory Committee of President Eisenhower, from November 22, 1957 (less than two months after the launch of _Sputnik_) to December 31, 1960, and also served under President Kennedy and President Johnson. The Eisenhower Library has many documents dealing with Purcell and these activities; many are still classified. He was also chairman of the Space Science and Technology Panel. Purcell's calculations about interstellar travel seem very simplistic for somebody with a proven record of creativity who is presumably very knowledgeable about space science and technology.

Impossibilists have raised other objections to manned space flight. These include that man would not be able to survive in a weightless environment, that meteors would puncture the spacecraft causing a loss of air, that the space radiation levels in the Van Allen belts and elsewhere would prevent people from surviving in space near the Earth or on the way to Mars or other stars for very long. These arguments rarely are quantitative. The anti-radiation people seem to forget that it is the radiation that reaches the astronaut that matters, not that which is in outer space. There are very energetic galactic rays that can smash cells, there are low-energy electrons that can't penetrate the spaceship skin, and others that get stopped by the space suit. People in the space program have worried about these items for 50 years. So far, they have not been show stoppers (nor have the meteors that come barreling by).

Of course, designers of nuclear submarines and aircraft carriers have been concerned with radiation shielding for many decades and have successfully built and operated such systems. One of the glaring errors of the impossibilists about deep space travel is that they seem so unaware of the large-scale nuclear (as opposed to chemical) systems that have been under development. For example, as noted in Chapter 1, the General Electric Company Aircraft Nuclear Propulsion (ANP) department in Evendale, Ohio, with help from the Oak Ridge National Laboratory and others, spent almost $1 billion on ANP. They actually operated jet engines on nuclear energy in the late in 1950s. General Electrics' budget for such work in 1958 alone was $100 million. It is useful to focus on the ignorance of nuclear projects by the impossibilists because, for space travel, weight is a major consideration. In World War II, the big bombs were 10-ton blockbusters. The energy released by H-bombs using nuclear fusion exploded by the United States and the Soviet Union is equivalent to as high as 57 million tons of explosives.

Several problems seem to constantly crop up amongst those who are skeptical of space travel: Impossibilists haven't done their homework about past work. They are not aware that technological progress comes from doing things differently in an unpredictable way. Moreover, they are unaware of the many large-scale programs conducted under security. They seem to believe that, because they are well-educated, they would be aware of all important research. In fact, they are clearly not aware, especially in areas away from their primary academic interest. In addition, they generally don't have a security clearance for advanced technology data. In short, it is certainly not scientific to do one's research by proclamation rather than investigation.

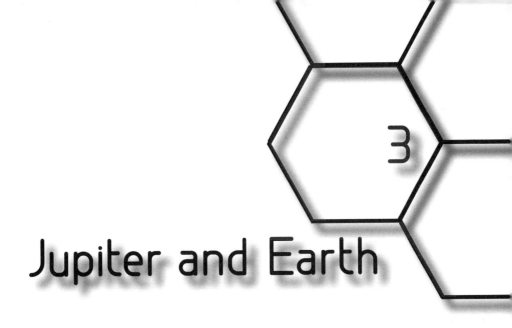

Jupiter and Earth

It is important to recognize that while astronomy is one of the oldest sciences, it also has a history of false and misleading claims based on how smart astronomers think they are and how nothing new can possibly be discovered that they don't already know. There is a very long list of absolutely false claims some astronomers have loudly made, such as Venus was only warm, Mars has always been dry, magnetic fields have no place in our understanding of the solar system environment, the composition of the stars will never be known, there is only one galaxy, and the stars are millions of years (rather than billions of years) old. This is one of the complications that interfere with a discussion of the role of magnetic fields. Almost everything we have learned about nuclear and sub-nuclear particles and ionizing radiation has been learned since 1890, and much after 1920.

Here are some dates to consider: X-rays were discovered in 1895. Radioactivity was discovered in 1896; the neutron in 1932. Fusion was formulated as the energy production mechanism in the stars in 1938; fission was discovered in 1939. The Van Allen Belts were discovered in 1958. The first radio signals received from space were picked up in the early 1930s by Karl Jansky at Bell Labs. Most of these discoveries were made entirely by accident and depended upon somebody being willing to say, "That is odd, let us look into it," rather than saying, "I must have made a mistake, since it makes no sense." For example, cosmic rays were discovered in 1912 by Austrian physicist Victor F. Hess. He was making measurements of radiation levels above the ground where radioactive materials in the ground provide more background radiation. Radioactivity had only been discovered, by accident, a few years earlier. Much to his surprise, he found that the intensity went up, rather than down, as he went to higher altitudes in his hot air balloon laboratory. He even was able to obtain data at an altitude of 17,500 feet. He, of course, had no idea that it would eventually be found that cosmic rays reaching Earth from the galaxy had energies as great as billions of electron volts. Normal chemical reactions involve a few electron volts. X-rays and gamma rays observed near the Earth and in nuclear reactors have energies of as much as 10 to 20 million electron volts. But some cosmic rays have energies as high as billions and billions of electron volts. Cosmic rays interact with particles and fields near the Earth to produce secondary particles, which in turn interact to produce other particles, and so on. As a general rule, the higher the energy the more penetrating the particle. It wasn't until 1936 that Hess was finally awarded the Nobel Prize in Physics for his discovery.

To illustrate how confident astronomers were that they had pretty much determined the characteristics of the universe in 1950, astronomer Fred Hoyle, who could not be considered very conservative, made the following comment in his book *The Nature of the Universe*: "Is it likely that any astonishing new developments are lying in wait for us? Is it possible that the cosmology of 500 years hence will extend as far beyond our present beliefs as our cosmology goes beyond that of Newton? I doubt whether this will be so. I am prepared to believe that there will be many advances in the detailed understanding of matters that still baffle us. But by and large, I think that our present picture will turn out to bear an approximate resemblance to the cosmologies of the future."[1]

Considering our new views about black holes, the Big Bang, worm holes, dark energy, string theory, nanotechnology, quasars, and lasers, it would seem that Hoyle was incorrect. Perhaps he could not have been expected to know in 1950 that literally billions of dollars would be spent on projects related to space and also the world of the small, or that entirely new capabilities for observation and computation would soon be developed. Computing went from desk calculators and slide rules to enormously fast computers with incredible data storage and handling capabilities. Simulations of possible stellar evolution that would have been totally impossible in 1950 can now be quickly run. Hoyle certainly didn't anticipate that billions of dollars would be spent on a host of new observatories operating in space above the limiting characteristics of the atmosphere. That would have seemed totally foolish. No planets in other solar systems had yet been discovered, nor were they anticipated. The real lesson that turns up over and over again is that technological progress comes from doing things differently in an unpredictable way.

There is no question that Dr. Hoyle would have been surprised at an August 14, 2009 release from NASA, in which it was announced that one of the last space shuttle missions in 2010 would deliver the truly extraordinary Alpha Magnetic Spectrometer to the International Space Station. It costs $1.5 billion and will include its own very sophisticated computer system. It will be searching for antimatter galaxies (if they exist), and for signs of dark energy (if it exists) and for strangelets, a theoretical form of matter containing strange quarks (if they exist). It will be looking for cosmic rays having energies up to 100 million terra-electron volts (TEV). (The collisions in the enormous Large Hadron Collider in Switzerland will only involve 7 TEV. 1 TEV is a million million electronvolts.) These are truly mind-boggling concepts.

Our local electromagnetic environment is sensitive to the electrons and protons and cosmic rays impinging on the Earth from not only the sun itself, but also from other portions of the galactic neighborhood. Cosmic rays are very powerful and influence the environment by ionizing the gases in the atmosphere and in the space between heavenly bodies. This makes them electrical conductors. One of the most important theoretical physics triumphs of the 20th century was the verification of the strange prediction that falls out of Einstein's general relativity that a large gravitational body, such as the sun, can bend light (electromagnetic radiation) passing by it on the way to the Earth. Major headlines touted the measurements made during an eclipse of the sun

Dr. Immanuel Velikovsky.
Courtesy of Frederic B. Jueneman.

in 1919: the apparent position of the stars whose light would pass near the sun on the way to Earth changed. Einstein's predictions were correct: the solar environment goes through cycles, as therefore does the production of solar storms, and thus particles, and thus the local environment. An additional focus, resulted from the discovery of the radiation belts known as the Van Allen belts existing at an altitude of hundreds to thousands of miles above the Earth. These are regions of high fluxes of charged particles. These offer protection, to some extent, against the external protons and electrons.

The electromagnetic environment and other surprising features in the solar system were among the major causes for vicious, unscientific attacks against Dr. Immanuel Velikovsky. His books, *Worlds in Collision*, *Ages in Chaos*, and *Earth in Upheaval*, were irrationally and publicly attacked by major figures such as Dr. Harlow Shapley, chairman of the Astronomy Department at Harvard, and another major Harvard astronomer, Cecilia Payne Gaposchkin. Their views on Velikovsky's books were published, even though they hadn't bothered to read his books because they "knew" his theories couldn't possibly be correct! "Science," as has often been the case, was wrong. Dr. Immanuel Velikovsky predicted that the atmosphere of Venus would be very hot, rather than less than 30 degrees, as was thought to be the case. Its surface is covered with clouds. He also predicted there would be radio waves from Jupiter. In an ironic twist of fate, astronomer Carl Sagan, who had vigorously attacked Velikovsky, was later given credit for finding that the atmosphere of Venus was hot!

In a remarkable letter to *Science*, the publication of the American Association for the Advancement of Science, Princeton University professor of physics Valentine Bargmann, and Columbia University astronomer Lloyd Motz courageously called attention to the fact that Velikovsky had correctly predicted both the presence of radio waves from Jupiter and the high temperature of Venus. They quoted Velikovsky, saying, on October 14, 1953, at Princeton, "The Planet Jupiter is cold, yet its gases are in motion. It appears probable to me that it sends out radio noises as do the sun and the stars. I suggest that this be investigated."[2] In June 1954, Velikovsky wrote his good friend Albert Einstein, asking him to use his influence to get Jupiter surveyed. That didn't happen. But in April 1955, B.F. Burke and K.L. Franklin[3] accidentally discovered and then carefully investigated strong radio signals from Jupiter. Because Jupiter was so cold, radio astronomers hadn't expected it to be emitting radio waves. Even more startling was the fact, determined by Radhakrishnan and Roberts[4] working at California Institute of Technology in 1960, that there was a radiation belt encompassing Jupiter giving 100 thousand billion times as much radio energy as the Van Allen belts around Earth.

In December 1956, Velikovsky had submitted a memorandum to H.H. Hess of the geology department at Princeton, suggesting the existence of a terrestrial magnetosphere reaching the moon—the magnetosphere was discovered by Van Allen in 1958. In 1950 Velikovsky stated that the surface of Venus must be very hot, contrary to the widely accepted temperatures of around 30 degrees—in 1961 it was finally realized, much to the astronomy community's chagrin, that the temperature was almost 600 degrees Kelvin.

Velikovsky was a true scholar trained in languages and in medicine. He decided to look to see if there were historical examples of catastrophes in other cultures besides those noted in the Old Testament—the flood, the plagues, the sun stopping at Jericho. He found many independent stories describing similar events in other old cultures. He found, because of his interest in Freud's work on Moses and Monotheism, that the dating of events in Egyptian history had to be changed. He actually had the bravery to suggest that Venus was a newcomer on the scene and that it and Mars had been at different distances from the sun than they are now, with Venus having been ripped out of Jupiter.

As might be expected, astronomers not only attacked Velikovsky, but also blackmailed MacMillan, the publisher of *Worlds in Collision*. If MacMillan refused to drop *Worlds in Collision*, they threatened to purchase their textbooks elsewhere. Even though the book was a runaway success on the top of the best-seller list, MacMillan sold the rights to Doubleday, which didn't have a textbook division, and then dismissed the editor responsible for getting the book published. This is hardly science at its finest. And, yes, as noted previously, radio waves *were* finally discovered coming from Jupiter.

Dr. Robert O. Becker.
Courtesy of SilverIons.org.

Astronomers have loudly insisted that there is no possible connection between the positions of the planets, and people and events on Earth. However, research has shown there *are* a number of connections not appreciated in the past. A man who made a major contribution to our understanding of the influence of the positions of the planets on the electric and magnetic field conditions on Earth was John H. Nelson, an electrical engineer. For many years, he had the unusual task of determining when RCA would have to use cables for short-wave radio signal transmission rather than broadcasting those signals. Cable utilization costs money. Nelson found that broadcast conditions could be predicted on the basis of planetary positions, amazing as that might sound. His success rate often ran as high as 90 percent, though he had to make predictions for four six-hour periods per day. He found that the transmissions were moderated by the heliocentric (relative to the sun) situation, especially with regard to the position of Jupiter.

Dr. Robert O. Becker (1923–2008), chief of the orthopaedic section at the Veterans Administration Hospital in Syracuse, New York, found important links between electric and magnetic fields and human behavior, during an incredible five decades of research. His very significant book is *The Body Electric: Electromagnetism and the Foundation of Life*[5] with Gary Selden. He began by investigating limb regeneration in amphibians (really quite remarkable, considering humans and many other animals have no such capability) and the healing of fractured bones in people. Some fractures seem unhealable, but in both situations he found, quite surprisingly, that properly administered electromagnetic fields could be of benefit, speeding up regeneration and fracture-healing. There were, as might be expected, many variables to be considered. He went where others had not gone before, was nominated twice for the Nobel Prize, and is considered the father of electromedicine and electrochemical cellular regeneration. He also had to fight against the powers that be—after all he was an orthopedic specialist, not a biochemist.

Dr. Becker also found surprising correlations between admissions to the psychiatric unit, and electric and magnetic fields. He had access to admission records for veterans and could try to correlate these with the electromagnetic environment—much to the surprise of some people.

Becker was very fortunate to get to know Dr. Howard Friedman, chief of psychology and clinical assistant professor at the Veterans Administration, to help in a project that grew out of Becker's watching the skies as an amateur during the International Geophysical Year (1957–1958). Observers received weekly reports of the state of the Earth's magnetic field. The question they explored was whether or not there was a correlation between perturbations in the electromagnetic field caused by solar storms and the rate of psychiatric admissions. They matched admissions of more than 28,000 patients at eight hospitals versus 67 magnetic storms throughout the previous four years. Significantly more patients were admitted to the psychiatric unit just after magnetic disturbances than when the field was stable.

In addition, consideration was given to patients already in the hospital. They selected 12 schizophrenics who were scheduled to remain in the hospital without any changes in treatment. Ward nurses provided a standard evaluation of their behavior after every eight-hour shift. The doctors then correlated the results with measurements of cosmic rays at government measuring stations in Colorado and Ontario, Canada. Magnetic fields are influenced by a number of factors, but cosmic ray intensities generally decrease after a solar storm. According to the reports of the nurses, there were various behavioral changes in almost all subjects one or two days after cosmic ray decreases. It was already known that one type of solar radiation, so called flares of low-energy cosmic rays from the sun, strongly perturbed the Earth's fields one or two days later.

Gradually there has come to be much concern about greatly increased electromagnetic exposures from power lines, microwaves, TV sets, computer monitors, and, of course, cell phones. Health-related studies have been done on electromagnetic effects as a function of the distance of a residence from power lines. People working in the electrical industry have been compared to people not working near electrical devices. There was a major battle between the standards-setting bodies in the United States and in the Soviet Union about maximum acceptable exposure to microwaves. The Unites States focused exclusively on levels high enough to heat tissue, but the Soviets claimed there were important effects at much lower levels. One concern was the exposure of employees working near major radar installations on ships as well as on land and in the sky. The United States has gradually come around to recognizing the possibility of non-thermal damage by microwaves.

As with almost all aspects of health, there is a marked variation of sensitivity to external stimuli. In the case of electromagnetic fields, it is not limited to the dose; also important can be the frequency and pulse width and other electrical characteristics. It often comes as a surprise to people that sensitivity to nuclear radiation varies greatly from person to person. This has become obvious when comparing people exposed to substantial radiation doses when being treated for cancer with, for example, the gamma rays from radioactive cobalt-60. Some can stand much higher doses without nausea and other side effects than can others.

It is commonplace for scientists to heap derision on anything that takes a serious look at the whole question of astrology. The opposition was formalized in a 1974 statement signed by 186 scientists, including 18 Nobel Prize–winners. It was widely distributed by the American Humanist Association and claimed there was nothing "scientific" to astrology. There were some scientists, such as Carl Sagan, who refused to sign it. Within two years the American Humanist Association was clearly shown to be guilty of scientific misconduct when it falsified data in an attempt to show that a test by French researchers, husband and wife Michel and Françoise Gaugelin, had negative results rather than the positive ones they had claimed. One wonders how many of the signers had actually known anything about the ancient practice of astrology.

The common notion is that being born during a particular sun sign is supposed to say something about one's character and capabilities. More serious astrologers take into account the position of many heavenly bodies at the time of birth. These include the moon and the sun, as well as the other planets of the solar system. Considering that various forms of astrology have been utilized for thousands of years, one might expect that there is some significance to it, even if there is no "scientific basis." The standard scientific reply is that the only connection between, say, the planet Jupiter and a newly born fetus is the very, very small gravitational effect of Jupiter on the infant, right? Not exactly.

As it happens, the positions of the planets do have another influence not normally considered: the planetary positions determine the center of gravity of the solar system. When, for example, Jupiter and Saturn are on the same side of the sun, the center of gravity is more on that side. Not too many years ago, the usual theory would have been that situation still has essentially no impact on babies being born on Earth. However, we have learned that changing the center of gravity of the solar system changes the conditions on the surface of the sun. In turn, these effects trigger or influence the flow of electrical particles (solar wind, etc.) impacting on the Earth. These, in turn, impact the electrical and magnetic fields on the Earth and in the atmosphere. Furthermore, at the moment of birth, the fetus is for the first time exposed to the electromagnetic environment surrounding the mother. (Amniotic fluid, which is similar to seawater, is an electrically conducting fluid that basically insulates the fetus against changes in the electromagnetic environment.)

Until we learned about electrical and magnetic fields within the solar system, around the planet, and in the atmosphere, we didn't even think of an electromagnetic effect on a human being. But we now know that solar particles produce the aurora borealis displays and can affect power grids on Earth, the operating capabilities of satellites in orbit, and the transmission of radio waves over long distances. Some very large-scale power outages have been produced by solar storm interactions with the power grid and inducing currents on the lines. Some nuclear weapons are designed to maximize the electromagnetic pulse (EMP) as a way of greatly impairing satellite communication systems, on which most military organizations on the planet are very much dependent.

In 2008, new research using a cluster of satellites showed that there are occasions (roughly every eight minutes) when magnetic portals are opened between the Earth and the sun during which Flux Transfer Events happen, pouring huge flows of charged particles through the temporary chutes to the Earth.[6] Not all the implications are understood, and the research is on-going.

Our knowledge of astrology goes back long before we knew anything about electricity. We take ready accessibility to electrical devices, such as TV sets, radios, computers, cell phones, lighting and propulsion systems, and much more, for granted now. It also hasn't been that long since we gained a good understanding of the electrical environment within the human body, though electrocardiograms and electroencephalograms are pretty much taken for granted today as ways to monitor the operating characteristics of the heart and the brain.

We have become aware that MRI devices are very useful in evaluating what is happening within the body in general and the brain in particular. The term "exploratory surgery" is not heard much anymore because of the magic offered by new technology to determine biological activities within the body. If we can determine activities within the body with electromagnetic devices, it doesn't sound unreasonable that electromagnetic fields might in turn have some influence on various internal organs, nerves, and so on. Furthermore, electromagnetic systems might be influenced by the conditions at the time of their first exposure to external magnetic and electric fields—a sort of preliminary tuning of the system. In other words, people born at the same time might be more similarly tuned than those born at different times.

As noted previously, the American Humanist Association published a kind of anti-astrology manifesto, "Objections to Astrology," in 1974 in the *American Humanist Association Journal*. The paper received a great deal of other press attention because copies were sent to every newspaper and magazine in the United States and Canada. After all, amongst the 186 signers were 18 Nobel Prize–winners. What followed was a pseudoscientific black mark on science perpetrated by those supposedly defending it against fraud and irrationality. The story is told in considerable detail by insider and skeptic Dennis Rawlins in a lengthy article in, of all places, *FATE Magazine*,[7] because there didn't seem to be any place else that would publish this horror story of vicious misrepresentation, false claims, and lies. The American Humanist Association was chaired by debunker Paul Kurtz,[8] who was excellent at self-promotion and directing Prometheus Press, which publishes almost entirely debunking books. He asked a skeptic, Marcello Truzzi, to become chairman of the newly formed Committee for the Scientific Investigation of the Claims of the Paranormal (CSICOP). This more recently morphed into the CSI, the Committee for Skeptical Investigation. The publication of this impressive-sounding organization is the *Skeptical Inquirer*. The mast-head lists many distinguished scientists, but the bulk of the scut work is done by strongly biased debunkers who are so convinced that such topics as ghosts, psychic phenomena, UFOs, and astrology are, by definition, false, and that they needn't be concerned about facts, evidence, and rational evaluations of counter-claims. Truzzi seemed much in favor of impartial inquiry. Kurtz was not. Carl Sagan, himself a skeptic, gave his reason for not signing the Manifesto as "Statements contradicting borderline, folk, or pseudoscience that appear to have an authoritarian tone can do more harm than good."[9] How right he was.

The article "Starbaby," written by CSICOP cofounder and associate editor of the *Skeptical Inquirer* Dennis Rawlins (its title a play on the Tarbaby of Br'er Rabbit stories), revealed the bias of the American Humanist Association's attacks on the Gaugelins. The brief summary at the beginning says, "They call themselves the Committee for the Scientific Investigation of Claims of the Paranormal. In fact they are a group of would-be debunkers who bungled their major investigation, falsified the results, covered up their errors and gave the boot to a colleague who threatened to tell the truth." The proximate cause of all this unethical activity was an article by the Gaugelins. They had shown that champion athletes were more likely to be born when Mars was at a certain location in the sky. They did a very detailed statistical comparison between the athletes and a comparable group of non-athletes. This didn't sit right with the debunkers.

Some of the debunkers were well known for their attacks on other areas of the paranormal. Perhaps the best known was Philip Klass. An illustration of his unscientific approach to the paranormal can be seen from his appearance on a Larry King show with Dr. David Jacobs, a Professor of History at Temple University, who was discussing his book about UFO abduction cases, *The Threat*,[10] which he had investigated in depth. Larry, at the beginning of the program, asked Klass what he thought of the book. Klass replied that all those people who believed they had been abducted had mental health problems. This was a very unscientific answer, because Klass had met none of the people and had no professional training or experience in psychology. (For scientific findings on the mental health of abductees, see Chapter 14 of *Science Was Wrong*.) His degree was in electrical engineering, and, for many years, he had been the senior avionics editor for *Aviation Week and Space Technology* magazine. In contrast, the foreword for Jacob's book had been written by Dr. John Mack, a professor of psychiatry at Harvard Medical School, who himself had been doing investigations of abduction cases. There was no question who was better suited for evaluating the mental health of the abductees. Later in the program, King asked Klass if he had actually read the book. Klass's answer was no. This illustrates the attitude of the attackers on the Gaugelins: "Don't bother me with the facts; my mind is made up."

TECHNOLOGY

Communication techniques didn't start changing until less than 200 years ago. When they began developing, the telegraph, telephone, television, Internet, and cell phones were all targets for the impossibilists. Sometimes it required true persistence from real scientists to overcome the inertia of the traditionalists and push forward in the field of communication.

The notion of cold fusion certainly upset a great many physicists who "knew" it was absolutely impossible for two chemists to produce energy in an inexpensive, low-temperature device, rather than the high-temperature, plasma, physics-type devices that had so far been unsuccessful despite the billions of dollars that had been spent on them. The furor subsided, and subsequent research by those unbowed by the derision seems to be bringing the new open-minded approach back into the spotlight, demonstrating the bias that had been shown in the past.

Communications

There is no question that communication techniques have advanced enormously in the last 170 years. The advancements were made in spite of the objections of very smart people who knew little about the particular techniques at hand. A good place to start is with the development of the telegraph in the 1830s by Samuel F.B. Morse, helped by Alfred Vail, in the development of the Morse code. An important factor when considering the difficulty of implementing their technology is that there were no electrical grids to provide the necessary electricity. New-fangled batteries were required. Wire had to be strung on wooden poles, which had to be provided and installed. Connections had to be made. It was indeed a new ball game. On May 24, l844, in an important event, the Supreme Court Chambers in Washington, D.C., sent the widely publicized message "What hath God wrought" to Baltimore on a demonstration line. By 1851, there were more than 50 separate telegraph companies in the United States. The reason for having so many companies is that the owners of the telegraph patents were unable to convince the United States and other governments of the usefulness of the invention. The government turned down an asking price of $100,000. So licenses were sold and many independent companies were established. Many companies were finally consolidated in 1856 into the Western Union Telegraph Company, which still exists today. It and other consolidated companies started working together (usually with each controlling a particular area of the country).

It may well be a very good thing that the government didn't take up Morse and his cohorts on their offer to sell. Governments don't have a good record of innovative and aggressive development, or operating efficiently. Companies have to make a profit, and so are more willing to take risks and to work hard, whereas governments have little incentive to keep costs down. Deficits seem to be the rule rather than the exception. Empire building, job protection, and campaign spending seem to be more the government way. Business doesn't have that luxury.

Until the widespread implementation of the telegraph, relatively slow mail delivery was the "fastest" way of communicating. An illustration of one of the important rules for the development of new technology, namely that progress comes from doing things differently in an unpredictable way, was provided by the simple fact that "rapid" communication (mail delivery) to the West Coast of the United States in the 1850s had been provided first by steamships going from New York to the Isthmus of Panama, switching to a canoe and a mule to get to the other side, and then another steamship up the West Coast. Total time typically was 22 days. The first breakthrough involved the Pony Express, which began in April 1860. A major organization was formed, consisting of 100 stations, between 400 and 500 horses, and 80 young, preferably lightweight riders who galloped from station to station. They covered about 250 miles in a day. The total time going westward across what was often difficult terrain was 10 days. Going eastward usually took 11.5 days. The Pony Express was certainly a great success, even if not very profitable, for the men who started the company. Amazingly, only one mail packet was lost. The service was closed in October 1861, not because it was a failure, but because the railroad had been put through, and the Pacific Telegraph Line, completed in October 1961, followed on the railroad right of way. Telegraph operators took the place of horsemen. Message transmittal time was just a few minutes, rather than many days. Cyrus Field had already laid the first trans-Atlantic telegraph cable in 1858. The world was getting smaller, thanks to technology, and in spite of those who said it couldn't be done.

The telegraph was a great boon for communications at a distance (so long as one had an operator at each end able to speedily transmit or "read" the dots and dashes of the Morse code). There was definitely a profit to be made by expanding the service, and there were three obvious (but very unusual for the times) goals for improvement. First, there was a need to transmit (broadcast) the signals without being tied to long wires and many poles. Second, the ability to transmit many signals at the same time was necessary. Third, it was thought to be useful if one could communicate voice messages, and perhaps music, rather than just dots and dashes created by clicking a key. A skilled operator might be able to transmit 40 words a minute using Morse code, but most humans would like to be able to use speech, which would not only be faster, but would also use all the subtleties of language and tone we take for granted when speaking to someone in the same room. So goals were defined for radio, which is wireless, as opposed to the telegraph, and for voice, which was of course the telephone. It was clear that the more people able to be connected, the more costs could be brought down and the more profit could be made. There were formidable difficulties in obtaining patents and defending them against those wanting to use other people's technology without paying royalties or other fees.

It may be difficult for people today to realize that there was no electricity grid to provide the power to send out signals over the telegraph wire. Batteries were the source of the electricity, once it was realized that electrostatic generators (such as moving leather against glass) were not terribly efficient. The major impetus for the development of better batteries was the work of Alessandro Volta. In 1800, he invented the voltaic pile. It had alternating discs of zinc and copper, with pieces of cardboard soaked in brine between the discs. Afterward, many improvements were made throughout the years. A French engineer, Georges Leclanche, patented a carbon–zinc wet-cell battery in 1866. By 1868, 20,000 of his cells were being used with telegraph equipment. In 1901, Thomas Edison invented the alkaline storage battery. The first commercial electrical power plant wasn't built until 1882, and the first long-distance, high-voltage power transmission line wasn't built until 1917. In fact, most of rural America didn't have access to electricity until President Roosevelt established the Rural Electrification Agency in 1936.

Alexander Graham Bell was not the first to try to construct or patent a telephone. He is usually given the credit, though a case can be made for Theodore Vail or Elisha Gray, and, even before them, to Johan Philip Reis, Emile Berliner, and Antonio Meucci. But history is almost always written by the winners. Bell did work hard on his ideas, and he knew a great deal about speech, having worked on helping the deaf and hard of hearing. Also, he was fortunate in that he had the financial backing of his future father-in-law, Thomas Watson. On March 10, 1876, he transmitted his famous message, "Mr. Watson, come here; I want to see you." It wasn't until January 25, 1915, that Bell made the first transcontinental call from New York (again to Thomas Watson) to San Francisco. It would be easy to forget the very substantial investment that had to be made after 1876 in infrastructure, considerable legal battles (especially once the original patents ran out), and the means for many smaller companies to use the transmission lines of Bell's companies. A very important development was the introduction of the three-element vacuum tube (audion) invented by Lee De Forest in 1906, which provided the amplification needed for signal transmission over long distances. American Telephone and Telegraph bought the rights to use this invention in telephony, which also turned out to be vital in microwave transmission, radio, television, and radar. Another important development was the laying of undersea cables so that telephone calls (as opposed to telegraph transmissions of Morse code) could be made among North America, Europe, and Asia.

Certainly the large expenditure on those telegraph cables was one of the reasons that the cable companies fought the seemingly absurd notion that signals could be sent across the ocean through the air without wires connecting the sender and receiver. A major objection was the idea that radio waves supposedly moved only in straight lines. Because the Earth is round, it would only be possible to transmit relatively short distances before the Earth's curvature got in the way or the signals went off into space—or so it was claimed. The major efforts to show that this wasn't true were made by Guglielmo Marconi, who had been running experiments in England at greater and greater distances between transmitters and receivers, and was certain that the curvature of the Earth would not be a problem. There were questions as to how high above the ground the transmitting antenna had to be and how high and large the receiving antenna had to be, as sometimes tethered balloons or kites were used. There were also such practical questions as to what frequencies were best. It was also being recognized that the time of day mattered, as it was subsequently discovered that transmission was much more successful at night than during the day. This was later determined to be because of properties of the ionosphere that had not yet been discovered. The big day came when signals (Morse code) were radioed from the transmitter in Poldhu, Cornwall, in England to Signal Hill near St. John's, Newfoundland, one of the nearest parts of North America to England, on December 12, 1901. Marconi and his assistant at Signal Hill claimed that they heard the signals three times at the times at which they were supposed to be sent. This was noted in Marconi's notebook. Ironically the instructions as to when to send were sent by undersea cable. The signal was three dots (the letter S). More than 100 years later, debates are still being held as to whether that reception was even possible and whether the clicks could be distinguished from the static.

Marconi did many follow-up experiments monitoring signals from ships at sea to get a handle on the distances over which signals could be transmitted. There was a lot of publicity about the successful experiment, and Marconi's company set up a business to establish transmitters and receivers on ships, supplying operators and equipment. The first known instance when a ship in distress sent wireless signals was in 1899, when a ship near the coast of England hit a lightship. Rescue ships responded swiftly to the emergency wireless telegraph call. The most famous example happened on April 15, 1912, when the "unsinkable" *Titanic* hit an iceberg in the North Atlantic. The radio operators transmitted positional data, and the ship *Carpathia* arrived soon enough to save about 700 people. Remember, this was Morse code wireless telegraphy, not voice transmission.

Heinrich Hertz had actually produced the first radio signals in 1886, based on the earlier theoretical work of Michael Faraday and Joseph Henry. In reality, there had been work done on voice transmission by radio not long thereafter. Reginald Fessenden had transmitted voices over short distances, and on December 24, 1906, he transmitted the first entertainment when he played the violin and he and his wife sang songs. Operators at sea all along the East Coast of the United States, who had been equipped with radio receivers and asked to listen, were surprised to pick up those broadcasts. However, as late as 1921, David Sarnoff, later head of RCA, stated, "The wireless music box has no imaginable commercial value. Who would pay for a message sent to no one in particular?"[1] It turned out that many people would.

It is not surprising that some of the initial responses to telephones were similar to those stated about radio, including asking, "What good would a telephone be?" At the beginning there was almost nobody to call; there were no telephone lines connecting each city household with the system—no switchboards, no operators, no phone directories. Constant improvements were being made as the number of phones grew. Income for the companies grew as well, so that more money was available to create the required infrastructure. Much effort went into the development of automation techniques, such as the dial, to reduce the number of operators required by the many systems. A major boon involved the invention of the transistor, which many consider the greatest innovation of the 20th century. The development was done, as might be expected, at the Bell Laboratory Division of American Telegraph and Telephone Company. The three inventors (all physicists) were Dr. William Shockley, Dr. Walter H. Brattain, and Dr. John Bardeen. The focus of the group was on the possible use of semiconductors to create a device that could replace the vacuum tube. Despite the success of De Forrest's audion and many other vacuum tubes, they took up a lot of space, were slow to warm up, and used a lot of energy, besides not being as dependable as was desired. It was late in 1947 when Bardeen and Brattain focused on the simple point-contact device that was able to control the flow of current through a circuit, at first using germanium and silicon crystalline devices. Shockley was the group leader, but did not much approve of the idea—Bardeen and Brattain mounted their experiments on a cart so as to be able to move it away when Shockley was present. Success was achieved on December 16, 1947. Shockley, being somewhat irked at not being part of the Bardeen and Brattain work, frantically worked out a number of theoretical problems (within the next few weeks) about how the new devices might work, and arrived at a somewhat different device, a junction transistor, by the end of January 1948. The first public announcement was not until the summer of 1948, because of patent considerations. The three were joint recipients of the Nobel Prize in Physics in 1956.

It took several years for manufacturing techniques to be worked out, using somewhat exotic materials and requiring very high standards of material purity and manufacturing technology. Perhaps surprisingly, only one of the three, Brattain, remained with Bell for very long. Shockley went to California in 1955 and started the Shockley Semiconductor Company of Beckman Instruments in Mountainview. This was really the beginning of the Silicon Valley agglomeration of high-tech companies. Bardeen became a professor of physics at the University of Illinois in 1951, continued physics research, and won a second Nobel Prize in 1972 for his work on superconductivity.

It took several years and much military support for the new transistor-based industry to be off and running, so that now transistors are used in a great variety of products and processes. Cost was more important in civilian applications, such as transistor radios, than for military applications where the transistors led to lighter weight, smaller, and lower-power-consumption devices that could be flown in airplanes. A major breakthrough occurred with the development of the integrated circuit (IC) in late 1958 by Jack Kilby at Texas Instruments, and a bit later by Ralph Noyce of Fairchild. It allowed many transistors to be placed on one small piece of silicon or germanium. Speeds of processing have constantly been increased, and size and power consumption reduced. Kilby received a Nobel Prize in Physics in 2000, one of the relatively small number given for work in applied physics—and "only" 42 years after his invention.

Within 50 years of the operation of the first transistor, the industry became a $150 billion industry, producing, in 1997, about 10,000 transistors per day for every human being on Earth. The transistor, and later the integrated circuit, were important in the development of another very important means of communication: television. As might be expected, in the early days, there were a number of negative comments about television from prominent people thought to be wise in the ways of technology.

In 1925, the editor of the *Daily Express* newspaper in London refused to see John Logie Baird, one of the first inventors of the television. As he told his associate, "For God's sake go down to reception and get rid of a lunatic who's down there. He says he's got a machine for seeing by wireless!

Dr. Jack Kilby. Courtesy of Texas Instruments Corporation.

Watch him. He may have a razor with him."[2] As it happens, Baird is indeed credited with providing the first television demonstration on January 26, 1926.

In 1926, Lee De Forest, who had invented the audion (electronic tube) so important to the development of radio, stated, "Theoretically and technically television may be feasible, commercially and financially I consider it an impossibility, a development of which we need waste little time dreaming."[3] Was he really in a position to make such a negative prediction? Or was he trying to preserve his industry—radio? About 30 years later, he was also claiming that man would never go to the moon, again without any indication of any relevant knowledge. Rex Lambert, editor of the London-based *Radio Times*, in an editorial in *The Listener* in 1936, was quoted as saying, "Television won't matter in your lifetime or mine."[4]

It is probably not surprising that Darryl Zanuck, head of 20th Century Fox Studios, in 1946, said, "Video won't be able to hold onto any market it captures after the first six months. People will soon get tired of staring at a plywood box every night."[5] Obviously, he would have preferred they go to the movies, his line of work. Now Hollywood movie studios do indeed provide many movies and other programming for television, in addition to the silver screen. Strangely, many families gathered in front of their radios at that time to listen to dramatic presentations, ongoing soap operas, sports events, news, and much more, "seeing" the programs in their heads, so one would think that the concept of the television would have been more widely accepted by those in the media. Another top-notch technologist, Thomas Alva Edison, reportedly said in 1922, "The radio craze will die out in time."[6] Others later on were equally convinced that the growth of TV would end radio broadcasting. It didn't.

Starting in 1950 or so, there was a slow but steady increase in the number of TV sets being produced. At the beginning, they all used vacuum tubes, and therefore were bulky and used a lot of energy. Initially, there were only very special broadcasts and small screens a few inches across. Many of the first sets were mostly mechanical, with rotating discs and such, as opposed to electronic. Gradually, networks started providing programming. The sale of advertising provided funds for expansion and much more varied programming. Between 1950 and 1960, about 60 million sets were produced in the United States. Eventually, cable and satellite television came along and boomed as well. It is not unusual now for one set to be able to receive 100 or more channels and for families to have several sets. But back then there was resistance. In 1961, T. Craven, FCC commissioner stated, "There is practically no chance communications space satellites will be used to provide better telephone, telegraph, television, or radio service inside the United States."[7] As it happens, the first commercial communications satellite went into service in 1965. He obviously was wrong.

It is of interest that the Space Age became an important part of the communications world despite, as noted in Chapter 2, the early resistance to space systems. One of the early arguments made vociferously was that if man were to go into space, it would be necessary to be in touch by radio. But radio stations use a tremendous amount of power, broadcasting at 5,000–100,000 watts, and the range is certainly limited. No satellite could lift such a heavy system providing that much power, was the argument. In actuality, small nuclear power plants have been launched for decades by the Russians—at least 35 at last count. Also very important are the *Pioneer 10* and *11* spacecraft. These are still sending signals back (from beyond the edge of the solar system) though launched about 30 years ago. The transmitters broadcast at 8 watts, the power used in a Christmas tree light bulb. However, unlike Earth-based stations, they were designed to send signals only in the direction of Earth, not in all directions. The receiving antenna was not a pair of rabbit ears, but a huge antenna such as the one at Goldstone near Barstow, California, which is 70 meters (230 feet) in diameter, and has enormous sensitivity. Furthermore, the power supply consisted of radioisotope thermoelectric generators, having no moving parts and using radioactive plutonium-238 as the heat source. It has a very long half-life indeed.

An important aspect in improving communications has been the introduction of optical fibers to replace wire. It didn't happen overnight, but has been the result of work by many companies and many government agencies. Fibers have several major advantages: They can conduct a signal for much greater distances than can wire without a need for amplification; they can handle far more data, and bandwidth has been crucial coupled with the development of lasers. (Both were developed for other reasons than to improve and extend the Internet, which in a very short time has become an enormously successful and widely distributed form of communication.)

The notion that many people would have a computer at home was ludicrous not too long ago. One reason given was that they would have to be enormous. *Popular Mechanics* in March 1949 stated, "Where a calculator on the Eniac is equipped with 18,000 vacuum tubes and weighs 30 tons, computers in the future may have only 1,000 vacuum tubes and weigh only 1.5 tons."[8] IBM seemed to think that the only possible uses for a home computer would be for storing recipes, keeping a checkbook, and compiling an address directory. Ken Olsen, president, chairman, and founder of Digital Equipment, claimed in 1977, "There is no reason for any individual to have a computer in their home."[9] Computers were assumed to be useful to large corporations and research groups such as the Los Alamos National Laboratory, for, for example, simulating the effects of nuclear weapons and using equipment leased from IBM for large-scale number crunching. To do this, in the late 1950s, an engineer would fill out input data sheets describing a particular system. The input data was keypunched. A box of computer cards was then carried with a big reel of magnetic tape having the program of interest to a special facility where a physically large computer was operated in a heavily air-conditioned facility. Printed output would be brought back to the user, who never went near the actual computer. In 1948, John Von Neumann, one of the top mathematicians in the world, stated, "We have reached the limits of what is possible with computers."[10] In the late 1950s, it cost about $500 per hour to use an IBM 704 computer with a memory of 64 kilobytes. As time went on, memories were being measured in megabytes, then gigabytes, and now terabytes. Personal computers usually have at least 160 gigabytes of storage space on a quite small hard disc. Now, quite sophisticated computers are available for under $1,000.

A terabyte is 1,000 gigabytes. A gigabyte is 1,000 megabytes. A megabyte is 1,000 kilobytes. A typical word consists of about six bytes of information.

The enormous advancements in the transistor and the integrated circuit, plus much competition, were essential to the development of two of today's most important means of communication: the Internet and the cell-phone. The Internet was conceived, originally in 1965, as a means of communication between various government research organizations focused on Cold War applications. Today it is a worldwide communications network consisting of thousands of networks, connected mostly by fiber optic cabling.

The primary early mover was the Defense Advanced Research Projects Agency (DARPA) in the middle of the Cold War in 1968. The telephone network served as an early model, but this was a major change, because the Internet used packet switching (independently invented by three research groups), as opposed to telephones, which used circuit switching. In 1971, there were 23 Internet host computers, mostly universities. By 1979 there were 111 hosts. Ten years later, there were an astonishing 100,000 hosts. And by 2009 the total number of Internet users worldwide was 1.73 billion, exploding from the 361 million users at the end of 2000.

A major factor in its growth has been the result of Moore's law, named after Gordon Moore of Intel, the major manufacturer of microchips. He predicted in 1965 that the processing speed and overall capabilities of new chips would grow by a factor of 2 every two years. He was right. Intel was a leader in improving the chips, putting more and more transistors on a smaller and smaller chip, which could operate faster and faster. (Their most recent integrated circuit has an astonishing two billion transistors.) One reason that IBM, accustomed to building huge mainframe computers, was slow to start in the personal computer business was a lack of imagination. An engineer in the Advanced Computing Systems Division at IBM commented in 1968 about the microchip, "But what the hell is it good for?"[11] IBM found out soon enough as personal computers became the rage and mainframe sales dropped precipitously.

Cell-phone growth has been equally impressive and again primarily because of rapid miniaturization in size and cost, and increases in both the efficiency and applications of microchips. In China and other parts of the world, more people have cell phones than have land lines. Inventor Martin Cooper of Motorola made the first recorded cell-phone call on April 3, 1973. It was to his competitor at AT&T—sort of a "gotcha" call. The phone weighed 30 ounces and was the size of a brick. By 1983 they had reduced the weight to 16 ounces. The cost was also quite hefty at $3,500 each. Today cell phones weigh a few ounces and are computers, cameras, phones, and Internet service devices all in one. Research in Motion has become one of the most successful companies in Canada because of its BlackBerry cellular devices.

It is worth noting that an important part of many communications devices, from printers to checkout counter devices, is the laser. It was invented in 1960 by independent groups in the United States and Russia. It covers an enormous range of utility from the tiny solid-state devices in CD players and the like, to systems capable of communicating to another similar system many light-years away, or, hopefully, inducing nuclear fusion. Its inventors, Theodore Maiman and Gordon Gould (it took Gould 27 years to get legal recognition from the patent office), were certainly not thinking at all about the practical application so often used today in such devices as CD players, DVD burners, barcode readers, surgical cutting devices, light pointers, and a myriad more. The energy range available runs from a few milliwatts to a million billion watts. The laser is a kind of direct descendent of the microwave maser, first demonstrated in 1953 by Charles Townes, though a number of eminent scientists said it was theoretically impossible. Two major areas of future interest are the use of lasers as part of the effort to detect signals from (and send signals to) extraterrestrial civilizations, and their use to induce nuclear fusion. Both involve very high power levels, as do the use of lasers to knock down aircraft and rockets.

It seems certain that new and improved techniques for communication will be developed in spite of resistance from arrogant experts thinking they know enough to say there will be no important changes or improvements. History has proven them wrong over and over again.

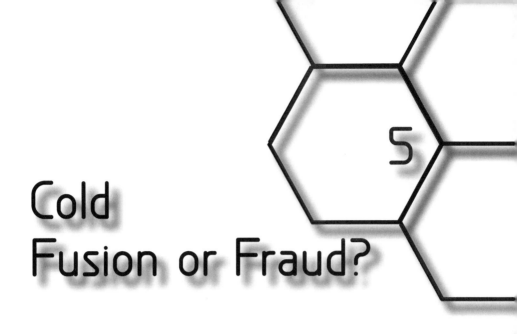

Cold Fusion or Fraud?

The announcement on March 23, 1989, of the discovery of a very exciting new form of energy production called "cold fusion" drew enormous attention from all over the world. The announcement was made by Stanley Pons and Martin Fleischmann in Salt Lake City. As shall be noted, there was some incredulity. After all, these men were both electrochemists, not nuclear physicists. They described a tabletop device that used electrodes made of the rare and expensive metal palladium, passing an electrical current through a bath of heavy water, but without producing a lot of radiation and without high temperatures. Both heat and radiation were thought to be prerequisites of a fusion system—in contrast to cold fusion, a great deal is known about "hot" fusion; after all, it powers the sun and all the other stars, as well as hydrogen bombs. And in both stars and hydrogen bombs, fusion involves very high temperatures and copious amounts of radiation.

Most people at the time were only vaguely aware of heavy water. Normal water is a molecule consisting of two hydrogen atoms and one oxygen atom, hence H_2O. Heavy water molecules consist of two deuterium atoms and one oxygen atom. The non-radioactive deuterium (heavy hydrogen) consists of a nucleus having a proton and a neutron and is about twice as heavy as a hydrogen atom, which has only a proton. Both hydrogen and deuterium have one electron in orbit around the nucleus; hence, their chemical properties are very similar, though their nuclear properties are not.

Dr. Stanley Pons and Dr. Martin Fleischmann. Courtesy of the University of Utah.

Only a tiny percentage of hydrogen atoms are actually deuterons. In seawater, for example, there is only one deuterium atom per 6,400 hydrogen atoms. Thus, unlike normal hydrogen, it is quite expensive to produce. Large chemical processing plants are required.

Because a proton has very close to the same mass as a neutron, when a neutron that is produced in nuclear fission strikes a hydrogen atom, for example, it is similar to a billiard ball striking another one: it can lose all its energy. It would take more collisions between the ball and, say, another ball twice as heavy, to lose its energy or slow down from the high speed it has in a fission nuclear reactor. The great majority of nuclear reactors in the United States use normal (light water) as their coolant and moderator. The moderator slows down the neutrons, which is necessary because fast neutrons are much less likely to be captured to produce fission by uranium-235 than are slow ones. Deuterium, in addition, is much less likely to capture a neutron than is hydrogen, so that the neutron may be scattered many times before it is captured by the U-235. The net upshot is that light-water cooled and moderated reactors require the use of enriched uranium having more than the normal amount of U-235 (2 or 3 percent, compared to the usual 0.7 percent). For nuclear weapons, one wants very highly enriched uranium. Thus, in most countries that have expensive enrichment plants to produce either U-235 or plutonium for weapons, they will have enriched fuel for power reactors. In countries not producing nuclear weapons, there is a great advantage in using heavy-water moderated and cooled reactors. All of Canada's reactors and those it has built in other countries are CANDU systems: CANadian Deuterium Uranium.

Nuclear fusion, as it is generally known, involves the combination of two light nuclei to make a heavier one, with a small amount of mass being converted to a huge amount of energy. Energy really does equal mass times the speed of light squared, which is a very big number indeed. Fission is more or less the reverse. A big U-235 nucleus absorbs a neutron and splits (fissions) into two or more smaller nuclei and again converts a little mass into a lot of energy. Whereas normal chemical reactions, such as burning gasoline, produce energies of a few electron volts per reaction, fission and fusion involve millions of electron volts per reaction. This is why there is so much interest in fission and fusion for energy production besides for use in weapons.

The isotopes involved in "normal" fusion are usually those of the very light elements, such as hydrogen, helium, and lithium, so there are many different combinations. Almost all the fusion reactions also produce neutrons and gamma rays. The major difficulty is that protons, deuterons, and lithium nuclei are all charged positively, having only neutral neutrons and positively charged protons in their nuclei. However, Mother Nature has provided a basic rule: positively charged particles repel each other. Thus, it normally takes a great deal of energy to get them close enough to overcome their repulsion and fuse.

Deuterium was discovered in 1931 by Dr. Harold C. Urey, for which he received the Nobel Prize in 1934. This discovery, followed in 1932 by the discovery of the neutron by James Chadwick, entirely changed the world of physics and chemistry. Science had been wrong in believing that the nucleus consisted only of protons and electrons. Chadwick received the Nobel Prize in 1935.

In H-bombs, a small atomic bomb is used to produce the very high energies and temperatures to get the fusion rolling. The palladium heavy-water cells did not produce very high temperatures, so fusion seemed completely improbable. Another possibility is to accelerate deuterons to high energy to cause fusion to occur when they impact on a deuterium- or tritium-loaded target. Tritium (T) is an even rarer isotope of hydrogen than deuterium. It has two neutrons and one proton. Hydrocarbons, such as crude oil, have lots of hydrogen. DT reactions are produced in small accelerators operated down oil-well bore holes to produce neutrons, which bounce off hydrogen in the surrounding soil, and give a strong clue as to how much oil is in the neighborhood.

Palladium is a quite rare, relatively heavy element of the platinum family. Its specific gravity (density) is about 12.02 grams per cubic centimeter (g/cc) as compared to lead's 11.35 g/cc. Its melting point is 1,552 degrees C, which is much higher than lead's 327. It is often used as a catalyst to expedite chemical reactions, such as in catalytic converters. One of its very important properties, in the context of the cold fusion controversy because of its molecular structure, is that it can absorb a great deal of hydrogen—about 900 times its own volume. It is quite expensive, often costing more than gold. A major question raised was, could deuterium nuclei be forced by pressures in the structure of the palladium to fuse and produce heat?

It took very little time after the announcement of cold fusion for the attacks on Pons and Fleischmann to begin. One of the biggest problems was that scientific papers on the subject had not yet been published in a refereed (peer reviewed) scientific journal. These normally are reviewed very carefully, especially when they make rather astonishing claims. Discovery and announcement by press release is very much frowned upon by the scientific community, and for good reason. Research by proclamation is hardly scientific, and often turns out to be proven false.

A second problem is that Pons and Fleischmann were electrochemists of good reputation, but not nuclear physicists. It should be noted that the two had coauthored 29 papers between 1985 and 1988 in reputable scientific journals, and that Fleischmann was president of the International Society of Electrochemists from 1970 to 1972. He was awarded the medal for electrochemistry and thermodynamics by the Royal Society of London in 1979 and was elected to the Fellowship of the Royal Society in 1986. In short, he had an outstanding professional reputation. But a number of nuclear establishments were immediately suspicious, especially in the absence of a paper to review. How could one try to duplicate the experiments producing unexpected heat without very careful descriptions of the equipment and measuring techniques used? A further problem was that even Pons and Fleischmann could not always reproduce their own results. The energy output of the small devices seemed to ebb and flow, and, in some cases, to continue after the current in the cell was turned off, which seemed very strange indeed.

It is perhaps not surprising that their results were not always the same. The long history of metallurgy makes it clear that a metal's properties are dependent upon many different factors, often not well understood. For example, how pure was the palladium? What impurities were present? How was the metal treated: rolling, annealing, heat treating in air or vacuum? How well-controlled was the electrical current in the cell? How pure was the heavy water? One of the by-products of fusion should be helium, but was there helium already present as an impurity? Did it escape from the tops of the electrodes?

Palladium is used in catalytic converters and other systems as a catalyst. Surface characteristics are very important. As an example, the United States Navy, in the 1950s, looked at the possibility of building nuclear submarine reactors using a liquid sodium alloy as the heat transfer agent (coolant) because it is better than water and would be used at lower pressure. Sometimes piping, in contact with the sodium, was highly corroded for reasons unknown. Sometimes, using supposedly identical materials, it was not. It took much effort to determine that the corrosion resistance of the metal was very much dependent on the oxygen content of the sodium. Unfortunately, at that time it was difficult to measure the concentration to better than perhaps 100 parts per million. In fact there was a big difference, as determined later, between 20 and 50 parts per million.

Alloys of even common metals, such as iron and aluminum, are quite sensitive to the amount and type of alloying agents and impurities. (Iron rusts, whereas stainless steel, which is primarily iron, doesn't.) A major reason for the successful development of large passenger planes, such as the 747, has been the development of very large and powerful jet engines with turbine blades operating at much higher temperatures and pressures than were thought possible not too many years ago. The turbine materials are not cheap, although they are alloys of relatively cheap iron and nickel, on which millions of dollars have been spent to improve their properties. Because of these improvements, engine thrust-to-weight ratios are much higher than was the case a few decades ago.

The nuclear world contains a great many examples in which small amounts of an element can greatly change a material's nuclear characteristics. Normal uranium consists of 99.3 percent uranium-238, but the great majority of fissions in normal reactors take place in the 0.7 percent that is uranium-235. The element boron consists of two isotopes, boron-10 and boron-11. The B-10 nucleus is 3,000 times more likely to capture a neutron than is B-11. So, a little goes a long way. The situation is further complicated by the fact that the probability of neutron capture is very much dependent on the energy of the neutron. The neutrons produced by the fission process have energies more than a million times as great as those just bouncing around the reactor. The probability of neutron capture typically goes up as the energy goes down, but can sometimes vary dramatically with a slight change of energy. Measuring neutron levels produced by cold fusion, if it exists, would not be easy. In other words, some scientific processes are much more complicated than might be expected.

Solid-state physics devices, such as transistors and chips in microcircuits, depend on small amounts of various doping agents to determine their electronic properties. One of the reasons for the high cost of chip facilities is the extraordinary measures that must be taken to control the cleanliness of the manufacturing environment.

The history of physics has many examples of surprises turning up. Many in the physics community, who rebelled against the ideas of Pons and Fleischmann, seemed to forget that. For example, in 1911, a Dutch physicist at Leiden University, Heike Kammerlinghe Onnes, discovered that the element mercury completely lost its resistance to the flow of an electrical current when cooled to near absolute zero (-273 Celsius or -459 Fahrenheit). He had earlier, in 1908, determined how to produce liquid helium in order to achieve very, very low temperatures. This was exciting, because it meant that one could pass a high current through an electrical conductor without heating the conductor and thus produce a strong magnetic field associated with the current. Onnes was measuring the resistivity of a very pure sample of mercury because, being a liquid at room temperature, it could be distilled to drive off impurities. Despite the fact that accurate methods of determining the impurity levels did not exist, it was known even then that impurities could affect resistivity. As the temperature dropped to 4.3 degrees Kelvin (4.3 degrees above absolute zero), the resistivity astonishingly dropped not just to a lower value, but to zero. Onnes received a Nobel Prize for his amazing discovery in 1913. Several other metals were also found to become superconducting at temperatures close to absolute zero, including lead. Some scientists have managed to keep a current flowing in a superconducting very cold lead ring for years. If there was any resistance, the current would have faded out. This came as a total surprise.

Unfortunately, it was also found that there was a critical current that, if exceeded, caused the material to become a normal conductor due to the magnetic field that was created. Thus there were two problems: the low temperature required and the low critical magnetic field. It was thought these stood in the way of practical applications. Another strange effect was noted in 1933 by Walter Meissner and Robert Ochsenfeld who discovered that a superconductor will repel a magnetic field. This "diamagnetism," as it was called, permits a magnet to actually be levitated above a superconducting device—very weird. Gradually the critical temperature increased to as much as 17.5 degrees above absolute zero. It took a long time (until the mid-1950s) for a reasonably successful theory to be worked out about superconductivity. John Bardeen, Leon Cooper, and John Schrieffer (BCS) of Bell Labs received a Nobel Prize in 1972 for their elegant theoretical work in 1957, which was complicated but worked to explain superconductivity for simple elements and alloys. Gradually it was found not to be adequate for complex materials. Another surprising theory was put forth by Cambridge graduate student Brian D. Josephson in 1962. He predicted, strange as it sounds, that electrical current would flow between two superconducting materials, even when they are separated by a non-superconductor or insulator. This was truly incredible. He received a share of the Nobel Prize in Physics in 1973 for his insights, as his prediction was verified. His idea plays an important role in certain very sensitive instruments.

It was also found that new superconducting materials were discovered that had much higher critical temperatures and currents than was thought possible. For example, the physics world was shocked when Alex Muller and George Bednorz, who worked for the IBM Research

Laboratory in Switzerland, discovered that a brittle ceramic material became superconducting at 30 degrees above absolute zero (-240C).
Ceramics were not supposed to be able to conduct electricity. They
received a Nobel Prize in 1987 after verification by scientists initially
very skeptical about their results. A new high critical temperature (the
temperature at which a material becomes super conducting) was announced in 1993. It was an extraordinary 138 degrees above absolute
zero. The strange material was made of mercury, thallium, barium,
calcium, copper, and oxygen. To say the least, such combinations were
not normally manufactured materials.

The superconducting magnetic materials in the MRI systems in
many hospitals may be made of various alloys such as niobium-tin or
niobium-titanium. They have to be kept very, very cold. Superconducting
magnets are also very important in nuclear accelerators, such as the
Large Hadron Collider at the CERN facility in Europe. They force the
very energetic, charged particles being accelerated to take a proper
path. Scientists had long assumed that ceramics, being insulators and
not conductors like metals, would never become superconducting. They
were wrong. Ideally one would want to use liquid nitrogen (melting
point a "warm" -210C) to cool the materials. It is much cheaper and
more available than liquid helium, which has a boiling point of -268.9C.

The maximum critical temperatures kept increasing. Sometimes
materials surprised their developers, proving that the BCS theory didn't
cover all situations as it had been thought it would. As early as 1961,
John Luce, head of research at Aerojet General Nucleonics in San Ramon,
California, and previously head of (hot) nuclear fusion research at Oak
Ridge National Laboratory, managed to get an Air Force contract to
look at a fusion propulsion system because of the comparatively high
critical temperatures and currents that had been found for niobium-tin
superconducting wire. The high field magnets would help confine the
fusioning plasma. Trips to the stars might eventually be feasible, as the
particles ejected in a fusion rocket would have 10 million times as much
energy per particle as those in a chemical rocket.

Much more recently, a discovery by Los Alamos scientists involving
nanomaterials indicated that the electrical current–carrying capacity
could be greatly increased. It should be noted that many engineers are
hopeful of using superconducting wires to reduce losses in long-distance

power transmission cables and in generators. Japanese scientists actually tested an electromagnetic submarine using superconducting magnets to provide the required high field magnets, which interacted with electrically conducting seawater to move a submarine silently.

The point of this long discussion of superconductivity is that many surprises were found all along the way in regard to the internal properties of all sorts of strange materials, despite arrogant thoughts that everything was already known. It may well be that the Pons and Fleischmann work has opened a window on another strange world, where we are still new-comers. It must be noted that despite the energetic antipathy of much of the physics and journalistic communities, the U.S. Navy, Japanese research labs, and the U.S. Electric Power Research Institute have supported cold fusion research, as they are interested in methods of producing energy.

Another example of surprising results, but with a very different reaction, also dealing with unusual properties of materials, was the rather exotic discovery, by Rudolf Ludwig Mossbauer in 1958, of a strange effect now known universally as the Mossbauer Effect. He had discovered recoil—free emission and absorption of gamma rays by nuclei—contrary to all expectations. If the absorber and emitter are embedded in a lattice (a solid), the recoil may actually be taken up by the entire solid, making the energy loss negligible. Strange as it may sound to somebody who is not a nuclear physicist, this effect has been used to make very accurate measurements of energy levels in nuclei. Areas in which it has made a major contribution include the study of the chemical consequences of nuclear decay, the study of the nature of magnetic interactions in alloys containing iron, and the study of the effects of high pressure on chemical properties of materials. The situation was much the opposite of Pons and Fleischmann. They were looking at the big picture of heat production in an unusual environment and involving many different parameters. Mossbauer zeroed in on a very specific effect and made a major contribution to physics. He was quickly awarded a Nobel Prize in 1961.

Thousands of papers about cold fusion have been published since 1989, but many in non-refereed journals because of strong resistance from the physics community. There has even been a strong movement to drop the term "cold fusion" to use the more descriptive, and less contentious, "low energy nuclear reactions."

Pons and Fleischmann were hounded from their jobs. As it happens, it was pressure from the University of Utah concerning prioritizing patent rights that led to the premature announcement in the first place. Another professor had been doing work on "muon catalyzed" fusion. (A muon is a subnuclear particle that seems, under special conditions, to also cause low energy fusion to take place.) Marc Plotkin of the Pure Energy Systems News Service summarized progress in the field in his March 2004 article "Cold Fusion Heating Up—Pending Review by U.S. Department of Energy." He noted that in *New Energy Times*, science journalists Steven Krivit and Nadine Winocur had just released a 50-page report on the current state of cold fusion. They claimed that almost 15,000 cold fusion experiments had been performed since 1989 and that, early on, experimental results were erratic and inconsistent, often with positive results occurring in only about 10 percent of the experiments. But successful replications began to be observed much more frequently. Five years earlier, the Fleischmann-Pons effect had been observed in only 45 percent of the experiments. By 2004, according to Krivit and Winocur, the effect has been reproduced at a rate of 83 percent. Supposedly, experimenters in Japan, Romania, and Russia reported a reproducibility rate of 100 percent. This new success is due, as might be expected, to better methods of measuring excess heat and detecting the signatures of nuclear reactions and adjusting some of the many parameters. For example, the ratio of deuterium to palladium in the cells had to be increased. The density of the electric current in the system had to be increased. A certain minimum threshold value needed to be achieved.

One of the leaders of the attack pack has been Dr. Robert L. Park, now Emeritus Professor of Physics at the University of Maryland–College Park. He has, for many years, been attacking numerous unconventional ideas, often doing his research by proclamation rather than investigation, as can be seen by a careful review of the claims in his book *Voodoo Science: The Road From Foolishness to Fraud.* He came up with the term "Voodoo Science" supposedly to cover "pseudoscience, pathological science, junk science, and fraudulent science." There is an entire chapter on cold fusion. Park's chapter on cold fusion was very successfully debunked by Dr. Eugene Mallove who had, prior to being murdered, been editor of *Infinite Energy*. Mallove, unlike Park, had been carefully reviewing the relevant literature for years. He had shown courage in retiring from MIT in 1989, as a protest against what he felt was their false representation of the results of their cold fusion experiments. Mallove noted that Park had previously claimed there was no cold fusion because no helium-4 was found. More than 10 years later, Park was still making the same claim, even though such data had been published in peer reviewed journals for years. Park, not surprisingly, as a debunker of many ideas he doesn't like, also attacked the people instead of the data. He states, "How, I wondered, could Pons and Fleischmann have been working on their cold fusion idea for five years, as they claimed, without going to the library to find out what was already known about hydrogen in metals?" As noted previously, Fleischmann was one of the world's leading electrochemists and wrote textbooks about hydrogen in metals. He certainly knew far more than Park did.

One might wonder, if Park doesn't get his information about cold fusion from the literature, what are his sources? Apparently a major source is Dr. Douglas Morrison of CERN, another fact-resistant debunker who goes to international meetings and then passes his misinformation on to Park. A very incisive review of Park's book, with a primary focus on cold fusion, was published by Eugene Mallove in *Infinite Energy Magazine* (March/April 2000). Park is *also* opposed to many things such as manned space flight and UFOs. A discussion of his false picture of the Roswell Incident is in *Flying Saucers and Science* (New Page Books, 2009). One of Park's comments in *Voodoo Science* should definitely be applied to himself: "While it never pays to underestimate the human capacity for self deception, they must at some point begin to realize that things are not behaving as they had supposed."

A particularly interesting and important paper was presented at the 237th National Meeting of the American Chemical Society, in Salt Lake City, Utah, on March 23, 2009, the 20th anniversary of the Pons and Fleischmann announcement, which was also in Salt Lake City. A coauthor of the paper was analytical chemist Dr. Pamela Mosier-Boss of the U.S. Navy's Space and Naval Warfare Systems Center of San Diego. She described careful measurements of the highly energetic neutrons produced in something called a LENR Device. This paper was one of 30 on the topic presented during a special four-day symposium on "New

Energy Technology." Critics had wanted "neutrons." She found the neutrons using a new and more sensitive measuring technique.

Dr. Park, and many others, must have been very surprised when *60 Minutes* did a quite favorable 12-minute piece on cold fusion, "Cold Fusion Is Hot Again," on April 19, 2009. The show had sent an investigative team to Israel to visit Energetics Technology, one of the leaders in the field of cold fusion. One of the team members was Dr. Rob Duncan, Vice Chancellor of Research at the University of Missouri, who was considered an expert on techniques for measuring energy. A major concern has always been that measurements of energy input and output in the cold fusion devices were often not being done accurately. Duncan started as a nonbeliever and spent two days asking questions and doing some of his own measurements. Duncan, to his surprise, admitted on *60 Minutes* that the company is indeed producing excess heat/energy. He stated that he had thought he would never say that. Also to its credit, *60 Minutes* took the time to interview Dr. Fleischmann in England. That the public is interested was clear from the fact that the next day the story was listed as one of the most viewed segments on the program's Website.

Several books have been published about cold fusion, most of them negative and by now out of date. One that should remain relevant for longer than usual because it is very detailed is *The Rebirth of Cold Fusion: Real Science, Real Hope, Real Energy,* by Steven B. Krivit (editor of *New Energy Times*) and Nadine Winocur. It was very enthusiastically reviewed in the Summer 2005 issue of *Journal of Scientific Exploration.* An enthusiastic blurb on the back of the book by Sir Arthur C. Clarke reads, "The neglect of cold fusion is one of the biggest scandals in the history of science. This book takes a fresh look at this still unresolved debate. An unbiased reader finishing this book will sense that something strange and wonderful is happening on the fringes of science." A particularly interesting comment also appears from Nobel Prize–winning physicist Dr. Brian Josephson: "*The Rebirth of Cold Fusion* gives much insight into how the 'due processes' of science came up with a decision that now appears to have been precisely the wrong one."

No one knows what the future of cold fusion controversy will be. We do know that much damage was done by baseless attacks not only directed at Pons and Fleischmann, but also to those who might have made major contributions to solving the world's energy crisis.

MEDICINE

Historically, when new paradigms have threatened existing dogma, those who clung to archaic ideology have worked to suppress emerging scientific ideas. Nowhere is this more apparent than in the history of medicine. When a mid-19th century Austrian physician advanced germ theory as the cause of childbed fever, impossibilists throughout Europe launched ad hominem attacks, suggesting that he practiced pseudoscience. Smallpox inoculations were met with bitter controversy and the lives of proponents were threatened. And though the impossibilists from pharmaceutical companies and blood banks reassured hemophiliacs that the chance of contracting AIDS was a million to one, thousands who listened died from AIDS. Thousands more are forced to live with the deadly retrovirus.

Politics, Personalities, and Childbed Fever

Born the fourth of seven children to a merchant family, Ignaz Philipp Semmelweis enjoyed a happy childhood in Buda, a prosperous Hungarian city separated from Pest by the Danube. His only disadvantage was a faulty bilingual education that failed to impart competence in communication skills. He spoke a Germanic dialect at home and didn't acquire a proficiency in Hungarian until secondary school. This was an obstacle that followed him throughout his life. Semmelweis began his college education at the University of Pest, but transferred to Vienna, Austria, and earned a doctorate in medicine in 1844. He applied for a fellowship in the study of pathology but it was awarded to another doctor. Consequently, he took specialized courses in obstetrics, and, four months later, earned a Master of Midwifery degree. His education continued in obstetrics and gynecology, and he received his doctorate in surgery in 1845.

In 1846, when he was appointed to a two-year term as assistant in obstetrics at the Allgemeine Krankenhaus (General Hospital) in Vienna, one in six expectant mothers admitted to the lying-in unit in Division I died within days of childbirth. During epidemic periods, the scourge of childbed fever, also known as puerperal fever, claimed 100 percent of the postpartum mothers in lying-in units in some European hospitals, leaving families devastated. Many others developed early symptoms of the disease but were somehow able to stave it off. Frequently, the new-born infants, whose mothers succumbed to childbed fever, developed hauntingly similar symptoms and quickly perished.

Childbed fever, caused by a streptococcus bacterium, occurred up to two weeks after childbirth, typically in the uterus or genital tract, and became a systemic infection within hours. Of course, in the mid-19th century there was no scientific evidence that the bacterium was real, but the symptoms that new mothers were developing in the lying-in units of hospitals all over Europe were tragically real. The first sign was fever above 100 degrees and loss of appetite, followed by chills and severe headache. Next, the victim's abdomen became swollen and pain-ful. The flow of blood and mucus from the uterus generally increased and took on a characteristically offensive odor. The simplest change of position caused excruciating pain. Urine became dark and foul-smelling, respiration and heart rate increased, and the production of milk by the mammary glands decreased. Near the end, the skin and fingernails turned blue. Despite the standard treatment of bloodletting and warm turpentine poultices, few women recovered from childbed fever. A mix-ture of calomel and opium eased the pain, but within days the victims slipped into delirium and coma, and then, certain death.

This was a time of growing industrialization in Europe and its cit-ies were burgeoning with an overflow of working poor. Throughout the continent, newly constructed hospitals were a symbol of the rise in the government's social consciousness toward society's underclass. Although the wealthier segments of society delivered babies at home, poverty, displacement, illegitimacy, and birth complications forced many wom-en into the cramped, often squalid conditions of the lying-in wards. Disease was rampant and the mortality rate averaged 25 to 30 percent.[1]

Low-income expectant mothers apprehensively entered Allgemeine Krankenhaus, fearing that they would never emerge alive. In exchange for free medical care and nutrition, they were forced to submit to exami-nation and study during labor and convalescence. This teaching hospital offered obstetrics and gynecology students, along with midwives, hands-on education under the guidance of trained physicians. The majority of women and infants survived the two-week recuperation period, but during childbed fever epidemics, the mortality rate soared.

The scourge of childbed fever had not always been as epidemic as Ignaz Semmelweis found it. Under the direction of its former professor of midwifery, Dr. Johann Boer, the lying-in unit at Allgemeine Krankenhaus was a humanitarian program characterized by cleanliness, gentleness of technique, and respect for patients' emotional needs. He strongly discouraged the use of forceps and other instruments and advocated the practice of natural childbirth, including proper nutrition, exercise, and fresh air. He refused to perform autopsies on his patients for clinical instruction, but instead used an anatomically correct wooden model to teach his students pelvic anatomy. Under his watch, the mortality rate from childbed fever was among the lowest in Europe (1.25 percent of 71,000 patients).[2]

In 1784, Austria's Joseph II built a modern 1,600-bed hospital, Allgemeine Krankenhaus, which had the largest obstetrical unit in the world, in keeping with his mother's wishes. He sought out Europe's finest scientists and physicians to bolster his programs. The lying-in unit at the hospital was divided into two units: one for teaching medical students and the other for midwives.

This fine tradition changed when Johann Klein replaced Dr. Boer in 1823. The new director's close ties to the political establishment secured his position at the obstetrical unit. Although he failed to keep abreast with new medical advances, his views were firmly planted in the zeitgeist of the old guard, and this was reflected in his procedural changes in the lying-in units, which he operated with military authority. He divided the department into two wards: obstetrical and midwifery. Klein required the use of cadavers for clinical instruction in the medical students' division, but not the midwifery division. Further, he ordered multiple internal obstetrical examinations on laboring mothers by inexperienced medical students in the obstetrical unit. These exams often caused tearing in the patient's delicate tissues. He also reinstituted the use of forceps (a tong-like instrument used for grasping, manipulating, or extracting the fetus).

Under Klein's tenure, with the increase of intrusive exams and the decrease of sanitary conditions, mortality rates in shot up dramatically, but only in the obstetrical Division I. The mortality rate remained at a stable 1.5 percent in the midwifery division between 1841 and 1846. However, it averaged 13 to 17 percent in the obstetrical division rising as high as 20 to 50 percent during epidemic periods.[3]

Upon graduation from medical school, Semmelweis registered for a position at the obstetrical division under Johann Klein. Nearly two years passed before a vacancy opened up, and he was appointed obstetrical assistant at the lying-in unit in Division I. In the interim, he became associated with leading young professors at the university, including Karl von Rokitanski, the medical school's enlightened director of pathological anatomy; the professor of legal medicine, Jakob K. Kollerscha; the clinician, Josef Skoda; and the dermatologist, Ferdinand von Hebra.[4] He gave Semmelweis free access to the autopsy room for dissection and study.

Semmelweis had ample opportunity to observe pathogenic changes in the cadavers of women and newborns from the hospital's obstetrical unit. His instruction in obstetrics and gynecology, under Rokitanski and others, imparted leading-edge knowledge that led to his doctorate in surgery in November 1845.

Under Rokitanski's supervision, Semmelweis learned to reject his European contemporaries' tradition of explaining pathological results in terms of clinical findings alone. He understood that European doctors interpreted the cause of death in terms of their theoretical knowledge and experience, even when it contradicted the evidence. Whereas many physicians limited their findings to a patient's specific organ, Rokitanski stressed the importance of identifying pathological patterns within several organs. The vast majority of physicians during the mid-19th century made diagnoses founded in speculation. Semmelweis practiced Rokitanski's careful method of scientific observation. Under Rokitanski, he learned that clinical medicine must be based upon an accurate knowledge of pathology bolstered by meticulous observation and detailed documentation—not speculation.

In March 1846, he received his long-awaited appointment to Division I under Klein's supervision. His duties included the instruction of medical students, assistance in surgical procedures, and clinical examinations. By then, his unit had the dubious distinction of having a nearly 25-percent mortality rate. This stood in stark contrast to Division II, the midwifery unit's 3 percent. Those who delivered at home or in the street, despite adverse conditions, had a far higher survival rate than those who passed through the doors of Division I. Distressed by the high death rate in his division, Semmelweis immersed himself in his search for an understanding of the pathology leading to childbed fever.

As soon as daylight ascended, Dr. Semmelweis led his students to the mortuary for instruction in the dissection of his deceased puerperal patients. Here they observed the pathogenic processes that caused the fatalities. They examined the thick pus that collected along the affected organs and smelled the characteristic foul odor that emanated from the gaping wounds caused by abscesses in the swollen vagina, uterus, and ovaries. Often the infection spread beyond the reproductive organs into the other tissues in the abdominal cavity, including the bladder and bowel. If the infection had entered the bloodstream, the entire chest cavity and even the joints might exhibit the purulent condition. Mysteriously, the deceased infants exhibited organic processes identical to their mothers.

Many theories had been proposed to explain childbed fever, but no one knew how to prevent it. Some physicians favored the milk-metastasis theory, founded in the idea that the milk of nursing mothers traveled along a duct between the top of the uterus and the nipple. Childbed fever, it was said, occurred when an obstruction in the duct forced the milk downward into the pelvic cavity, and later, to various parts of the body. The theory's proponents believed that the foul-smelling, thick pus that collected throughout the deceased body was nothing more than clotted milk, and the white fluid represented the collection of liquid milk. Advocates of the milk-metastasis theory believed that because an inherent defect in the mother's physiology caused the disease, there was no need to search for a cure. It was an excellent example of the "blame the patient" mentality that pervaded much of Europe in the mid-19th century. It is difficult to comprehend how this ridiculous theory sustained support throughout the 17th, 18th, and 19th centuries, but it contributed to the deaths of thousands of postpartum mothers.

Other leading obstetricians hypothesized that the normal discharge of blood, tissue, and mucous from the vagina after childbirth somehow reversed direction, putrefied, and oozed upward into the affected woman's blood, organs, and tissues. Physicians postulated that the causes of lochia suppression could be found in cold feet, the intake of cold water, narrowness of the blood vessels, or the thickness of blood within the vessels.

Still other theorists advanced the age-old miasma theory that noxious atmosphere arising from putrid matter, including swamps, decaying animals and plants, and feces caused the disease. This theory was supported by the fact that lying-in units produced such a stench from feces, blood, bodily fluids, and infection that they had to be separated from the main hospital to prevent the spread of disease. The filthy, weeks-old sheets and linens contributed to the noxious environment. Weather conditions and solar and magnetic influences were also suspected causes, along with other silly notions such as the belief that a woman's fear of childbirth or going to the hospital caused the disease.

Dr. Klein, the Division I director, was steadfast in his contention that childbed fever could not be attributed to a previously undiscovered cause, and when entire rows of patients died within days of each other, he threw up his hands in the belief that he was not only powerless to stop the spread of disease, but blameless as well. He remained steadfast in the belief that the low mortality rate in Division II, the unit maintained by midwives, was merely coincidental. Likewise, he believed the high death rate among postpartum mothers in his unit was beyond his control. Locked in denial, Klein perpetuated the conditions that killed thousands of his patients.

Deeply disturbed by the intense suffering and certain death of the mothers in his care, Semmelweis was determined to reduce mortality rates in Division I. Based upon theories being taught at the Vienna Medical School, he instituted measures designed to prevent the disease. For instance, he ordered adequate ventilation and changed birthing procedures to reduce trauma during childbirth. All of his efforts were ineffective; mortality rates remained high.

The dictatorial Dr. Klein viewed Semmelweis's actions as a challenge to his authority. He was convinced that preventative measures were impossible in the face of an epidemic for which there was no cure. In his opinion, the young doctor's experimental changes were futile. Undeterred, Semmelweis tirelessly continued his research, which Klein viewed as a clear case of insubordination.

In 1847, a fatal accident occurred in the pathology laboratory that finally solved the mystery of childbed fever. Ignaz Semmelweis's good friend, Jacob Kolletscha, a professor of forensic medicine, became ill and died after his finger was accidently punctured with a scalpel during the dissection of an infected patient. Kolletscha's autopsy revealed pathological changes identical to those Semmelweis had been observing in the cadavers of childbed fever victims. Empirical observation implied that an infectious agent was inducing sepsis in the affected postpartum mothers and their infants. He deduced that the disease was a form of blood poisoning. The mid-19th century notion that microscopic pathogenic organisms resulted in rampant contamination flew in the face of the established theory and practice, in which merely wiping one's hands on a dirty apron was considered good hygiene.

Semmelweis suspected that he and his students were directly transferring pathogenic contamination from the autopsy room to their patients in the Division I lying-in unit. They spent the afternoon performing at least five pelvic examinations on every laboring mother. He reasoned that the septic particles on their unsanitary hands transferred directly to their patients' genital organs, and later entered their bloodstreams. In turn, the infection was passed to infants during the birthing process.

Semmelweis immediately began experimenting with various cleansing solutions. In May 1847, he introduced hand washing using a chlorinated lime solution and a nail brush prior to each student's entry into Division I. Later, after an entire row of patients died within days of each other, he ordered hand washing before each vaginal examination. Additionally, he demanded the changing of bed sheets more often than once a month. As a result, the mortality rate from childbed fever in Division I fell from 18 percent in May 1847 to less than 3 percent in the next six months. His careful observation and documentation laid the foundation in support of his contagion theory.

The following year, Semmelweis initiated the policy of isolating patients with infectious wounds and sanitizing obstetrical instruments such as forceps. Nevertheless, his success was met with skepticism and ridicule. Many of the older faculty members, including Dr. Klein, rejected the contagion theory and were resistant to Semmelweis's sanitation measures, speculating that the dramatic drop in mortality rates signaled the end of an epidemic. To Klein, the idea that the hands of a doctor, which were meant to cure disease, could infect his patients with a fatal illness was a major insult; to Semmelweis, the idea was a tragedy. The conservative Dr. Klein frowned upon Semmelweis's innovations and created additional obstacles to his professional development. At the end of his two-year assignment, Klein, an arch conservative, refused to reappoint Semmelweis and replaced him with his former obstetrical assistant, Carl Braun, who adhered to his own archaic belief system. The sanitation and hygiene measures that Semmelweis had instituted were abolished. Predictably, childbed fever mortality rates shot upward because inflated egos, obstinacy, and ignorance prevailed in the medical establishment.

With heightened resolve, Semmelweis applied for a position as instructor in obstetrics in midwifery. In the interim, he received a grant through the Vienna Academy of Sciences to conduct experiments to test his clinical findings. He designed a series of experiments in which he introduced putrid discharge from the cadavers of women who had died from childbed fever into the genital canals of puerperal rabbits. As predicted, all of the rabbits died from symptoms identical to childbed fever. The subsequent autopsies supported Semmelweis's hypothesis. His successive experiments included the injection of various infectious and noninfectious particles into the genital canals of rabbits. He concluded that when rabbits were injected with particulate matter from childbed fever victims they died from childbed fever. When they were injected with anything else, they remained healthy or developed milder infections.

Semmelweis had garnered the support of three influential faculty members: Rokitanski, Ferdinand Hebra, often considered the father of dermatology, and Josef Skoda, who introduced percussion, or tapping, to interpret clinical entities and pathological processes. They urged him to publish his findings, but Semmelweis was reluctant. He remained challenged by his inadequate early education. Hebra assisted Semmelweis by publishing the first accounts of his sanitation discovery in the journal of the Medical Society of Vienna, in December 1847 and April 1848. In October 1849, Josef Skoda supported Semmelweis's findings in an address to the Imperial and Royal Academy of Sciences. But both misrepresented some of the facts in Semmelweis's principles. This led to confusion and resistance within the medical community. Although his colleagues urged Semmelweis to publish his findings and speak to medical groups, he remained reluctant. However, after much persuasion, on May 15 and June 18, 1850, he presented his doctrine entitled "The Origin of Puerperal Fever" to the Association of Physicians in Vienna, of which Rokitanski was president. He corrected the mistakes that Hebra and Skoda made, but this did not temper the derision of his adversaries who didn't support his theory's basic tenets.

Dr. Eduard Lumpe, a member of the Royal Academy, countered Semmelweis with the argument that childbed fever mortality rates increased during the summer months but declined in the winter. Therefore, he reasoned, it must be attributed to other causes. It didn't occur to him that medical students performed dissection in the afternoon during short sunlight seasons, and in the morning when natural light was abundant. It stands to reason that puerperal fever was rampant when dissection occurred in the morning, followed by internal exams in Division I.

One of Semmelweis's greatest detractors was Friedrich Scanzoni, a recently appointed lecturer at the Prague General Hospital, who later became a renowned professor of obstetrics. He was offended by Semmelweis's presentation of critical reports about the mortality rate and septic conditions at the Vienna General Hospital. For this, Scanzoni launched a series of personal and philosophical attacks upon Semmelweis that would follow him throughout his career.

Ignaz Philipp Semmelweis postage stamp. Courtesy of the Republic of Austria Postal Service.

Finally, in March 1850, Semmelweis's application for the position of instructor of obstetrics in midwifery was approved, and he was appointed to begin instruction on October 10. The faculty had petitioned the Ministry of Education, at his request, to exempt Semmelweis from the midwifery policy that limited instructional activities to the use of an anatomically correct wooden mannequin. The Ministry granted the request, permitting him to instruct his students through the dissection of cadavers. But as if to rebuke Semmelweis for his non-compliance with the zeitgeist of the old guard, in October 1850, the terms of the agreement were revoked, and he was denied the right to teach from the cadaver. One can imagine the turmoil and sense of persecution experienced by Semmelweis. Five days later, in humiliation and defeat, he abruptly left Vienna and relocated to Pest without even saying good-bye to his friends and supporters. This caused a permanent rift between Semmelweis and some of his cohorts.

The failed Hungarian Revolution had left Pest in shambles. Semmelweis entered an archaic, politically and scientifically depressed environment.[5] However, on May 20, 1851, his spirits were lifted when he was appointed honorary senior physician in the obstetric division at St. Rochus Hospital in Pest. He found the hospital in the midst of an outbreak of childbed fever for which he immediately advocated the use of his preventative measures. Soon, the mortality rate dropped from 10 to 15 percent to .85 percent. In July 1855, he was appointed Professor of Theoretical and Practical Midwifery at the University of Pest, where he worked immediately to upgrade the once deplorable conditions in the obstetric division. Although his ideas initially met some ideological resistance coupled with severe economic restrictions, he began to gain respect and acceptance in Hungary. Sometimes his efforts were sabotaged by opponents, many of whom he alienated through sharp criticism of their methodology and ideas. For example, bed sheets and linens that contained putrid contaminants were left unchanged by a director of midwifery who opposed Semmelweis's attempts to enforce stricter sanitation measures. Additional opponents refused to wash their hands when Semmelweis was not there to enforce the policy. In each case, he documented a spike in the mortality rate. As a result, his ideas eventually took hold.

While he toiled in the maternity ward at St. Rochus Hospital and built a prosperous medical practice, his adversaries published scathing attacks upon his work. The inaccurate and misleading information in their articles led to false assumptions about his theory, and no amount of protestation by his advocates could alter their statements. The medical establishment, throughout Europe, worked tirelessly to promote its own archaic ideas, while declaring Semmelweis a charlatan. This intransigent ignorance led to the unnecessary deaths of thousands of women and infants.

Finally, his friend Markusovszky, editor of a new weekly medical journal *Orvosi Hetilap*, persuaded Semmelweis to once again publish his principles and guidelines. At Markusovzky's urging, he addressed the Medical Society of Pest-Buda on January 2, 1858. It was the first of four lectures that year that were subsequently published in *Orvosi Hetilap*. Later, Markusovszky wrote of Semmelweis, "Semmelweis expounded his teaching with the conviction before our society that only a man is capable of possessing who not only can fight for its truth but vouches for it with his life. His dedication to his work was evident at the meeting of the Medical Society and it deeply moved all those present."[6] Motivated by Markusovzky's praise and encouragement, in 1861, Semmelweis published his theory in a 543-page treatise *Die Aetiologie, der Begriff und die Prophylaxis des Kindbettfiebers* (*The Etiology, the Concept and the Prophylaxis of Childbed Fever*). The book, though somewhat poorly written, is said to be one of the best documented works in the history of medical publications. In section one, Semmelweis presented his historical, clinical, pathological, and experimental findings,

After the publication of his book, *The Etiology, the Concept and the Prophylaxis of Childbed Fever,* some obstetrical units adopted Semmelweis's sanitation methods, but, due to inadequate supervision, many were ineffective. supported by innumerable statistics he had presented in his first thesis "The Origin of Puerperal Fever." In the second section, he refuted the antiquated theories of his detractors and launched a passionate defense of his findings. He distributed his work widely to obstetricians and medical societies throughout Europe, but he received little response for his effort. His adversaries launched new attacks upon his findings, again misquoting and misrepresenting him. Other scientists, assuming their skeptical colleagues' information was accurate, made no changes in their medical hygiene practices.

In 1861 and 1862, he launched a series of merciless counter-attacks upon his critics in open letters. In a scathing letter to his nemesis Scanzoni, now a professor of obstetrics at Wurzburg, he wrote, "Should you, however, Herr Hofrath, without having disproved my doctrine, persist in writing and permitting to be written about epidemic childbed fever—, I declare that before God and the world that you are a murderer and the *History of Childbed Fever* would not be unjust to you if it memorialized you as a medical Nero, in payment for having been the first to set himself against my life-saving theory."[7] Despite his courageous struggle, Semmelweis failed to gain widespread acceptance for his contagion theory. Thereafter, his health began to decline.

Early in 1865, at age 47, it was becoming increasingly apparent to his wife and friends that Semmelweis was becoming mentally incompetent. His once robust body had withered to that of a frail, old man, and he exhibited the symptoms of an organic brain disorder. His mood became increasingly bellicose and irrational, leading to his admission to a Viennese psychiatric hospital that summer. In what can only be described as an ironic twist of fate, he died two weeks later from septicemia. When his chest cavity was opened, it revealed a large abscess that encompassed the thoracic cavity, including the pericardium and heart. Although the nature of his illness and cause of death is still debated, a 1963 forensic examination, during his disinterment and reburial in Budapest, revealed evidence of injuries suggestive of a brutal assault. Following his admission to the psychiatric hospital, unknown persons inflicted devastating injuries upon his left hand, four fingers of his right hand, arms, and chest.[8]

In death, with the exception of a few supporters, Semmelweis's ideas drifted into obscurity in the Germanic States and Europe. They did not gain wide acceptance until 1875, after Joseph Lister introduced antiseptic measures into surgery based on the works of germ theorist Louis Pasteur. By the end of the 19th century, antiseptic medical procedures similar to those advocated by Semmelweis were practiced worldwide.

It was obstinacy and ignorance combined with the blind acceptance of ideas promoted by the medical establishment that delayed the acceptance of the cause of childbed fever and preventive antiseptic and aseptic measures. Doctors were unwilling to consider the possibility that their hands needlessly caused the deaths of countless women and infants. Politics and personalities trumped innovative, scientifically sound measures. Hand washing was inexpensive and easy to implement, but doctors forbade it.

In 1877, the Prussian Minister of Culture, Education, and Medical Affairs introduced strict nationwide measures for the prevention of childbed fever. By the end of the 19th century, virtually all obstetrical units in the Germanic States complied with the aseptic teachings of Ignaz Philipp Semmelweis. In 1891, the University of Budapest appointed a memorial committee to honor the man who had become a national hero. In 1906, a memorial to Semmelweis was erected in Budapest, and in 1971, Heidelberg unveiled a memorial in his honor. The Medical University of Budapest where he worked is named after him. And finally, the Republic of Austria has honored him with a commemorative stamp. The tragedy lies in the fact that these honors were bestowed upon Semmelweis years after his struggle to save the lives of women and infants went down in defeat. Science was wrong.

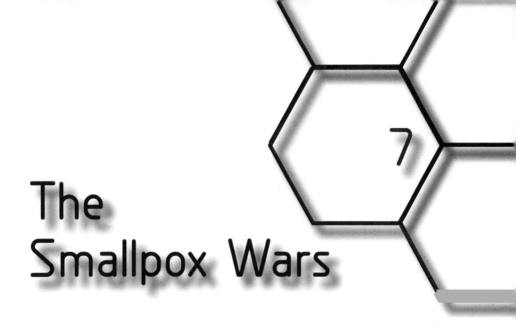

The Smallpox Wars

Smallpox is the most devastating pandemic ever to afflict the Earth's human population. It once spread indiscriminately through respiratory emissions, by skin-to-skin contact, and on contaminated clothing and bedding. There was no cure. It struck down people in every age group and social strata, taking a particularly heavy toll on children and young adults. In 18th-century Europe, 400,000 people died annually from smallpox. One-third of the survivors went blind. Not only did it cut a fatal path through its victims, its survivors were left with disfiguring scars. Beautiful women developed monstrous visages and children bore its lifelong pockmarks. It appeared suddenly, as a raging fever, headache, muscle pain, nausea, and vomiting. Next, pustules appeared on the face and inside the mouth, nose, and eyes, leading to blindness in a third of its victims. Finally, painful lesions blanketed the entire body, sometimes numbering in the thousands. Survivors generally recovered within a month's time, but forever carried the hideous evidence of their misfortune.

As early as 10000 BC, the first outbreaks of smallpox occurred in northeastern Africa and what is today Eastern Syria, Turkey, and Iraq.

The development of agriculture brought large numbers of people in close proximity to one another, facilitating human-to-human transmission. By 3000 BC, it swept into China and India. In 1157 BC, the Egyptian Pharaoh Ramses V succumbed to the disease for which there was no cure. In 490 BC, it devastated Athens, killing a third of the population. Near the end of the 6th century AD, smallpox appeared in Japan and Korea. It proliferated throughout Europe and Africa by AD 1000, and 500 years later, it carved out the history of the New World.

In 1518, Spanish sailors, bent on colonizing the New World and enslaving the natives, unknowingly delivered smallpox to the island of Hispaniola. It quickly spread to the indigenous population. Because they had never been exposed to smallpox, it killed nearly all that it touched. This depletion of the native slave labor forced the Spanish conquerors to import slaves from West Africa, where smallpox was endemic. The African slaves carried the virus to a larger percentage of the native inhabitants, decimating half. The highly contagious scourge, for which there was no effective treatment, spread rapidly throughout Central America, killing hundreds of thousands.

By 1520, Spanish conquistador Hernando Cortez faced certain defeat in his quest for slaves and gold in the New World. Overpowered by thousands of Aztec warriors in Mexico, his mission was all but doomed. However, in an uncanny twist of fate, an African slave brought to the New World by Cortez developed smallpox. It soon broke out in Tenochtitlan (modern-day Mexico City), and claimed the lives of half its native population, including Aztec Emperor Montezuma's brother Cuitlahuac, who had been ruling in his brother's absence. With Montezuma in Spanish custody and a multitude of senior Aztec leaders and warriors dead from smallpox, on August 13, 1521, a comparatively small band of Spanish troops entered the Aztec capital. They encountered such a horrendous scene of victims in the throes of death and decomposing corpses that they had to abandon the city until the intolerable stench had subsided. Without resistance, Cortez and a small ragtag army conquered the Aztec Empire, enslaved its people, and plundered its natural resources. Smallpox was the one deciding factor that brought down an entire empire.

In 1525, smallpox had spread to the Incan Empire in present-day Peru, killing Emperor Huayna Capac, his heir to the throne, and most of his senior military officers. A war of succession broke out when a battle for authority erupted between the legitimate heir and his illegitimate brother, resulting in a bloody five-year civil war. The opposing armies carried smallpox throughout the Incan Empire. It killed almost everyone in its path. This opened the door to Spanish conquistador Francisco Pizarro, who conquered the once mighty empire in 1532, with only 167 soldiers.

In 1614, a British ship landed on the Massachusetts coast and captured 24 Native Americans as slaves. It left behind a horrific smallpox epidemic. From 1614 to 1619 the disease decimated 90 to 96 percent of the indigenous population along the Massachusetts coast. When Miles Standish, the military commander elected by the Pilgrims for Plymouth Colony, arrived in 1620, he found few survivors.

When John Winthrop, a pilgrim leader and founder of the Massachusetts Bay Colony, arrived on the Mayflower in 1620, he attributed the carnage to an act of God. His naiveté about disease transmission is apparent in the following quote from a letter he wrote to his friend in England: "But for the natives in these parts, God hath so pursued them, as for 300 miles space the greatest part of them are swept away by smallpox which still continues among them. So as God has thereby cleared our title to this place, those who remain in these parts being in all not 50, have put themselves under our protection."[1] Smallpox laid fallow another land and its inhabitants to foreign invaders.

Although smallpox accidently spread to native populations throughout the Americas, it was also used in warfare. As we will see, military officers intentionally infected their enemy in order to expedite their conquest. During the French and Indian War (1754–1763), British forces under the command of Lord Jeffrey Amherst conquered French and Indian troops in Quebec City, Montreal, and Louisbourg, Nova Scotia. This victory earned him the title of British Governor General in the Canadian territories. In 1763, a siege upon Fort Pitt by Ottawa Chief Pontiac's warriors angered Amherst. The native warriors felt justified in their attack because, unlike the French, Amherst refused to offer provisions to their people in exchange for friendship. The Governor General sought retribution against the native warriors. Smallpox had broken out at Fort Pitt, and in a clear plan to inflict biological warfare upon Pontiac's nation, Colonel Henry Bouquet suggested that Amherst distribute blankets to the Ottawa tribe "to innoculate the Indians."[2] Amherst bestowed the gift of blankets and cloths insufflated with smallpox scabs to the tribe, and urged Bouquet "to try Every other method that can serve to Extirpate this Execrable Race."[3] To hasten the elimination of the native inhabitants, Amherst wrote to the Superintendent of the Northern Indian Department, "...Measures to be taken as would Bring the Total Extirpation of those Indian Nations."[4] The insidious intent to launch the world's most lethal biological weapon was carried out effectively, killing thousands. Thereafter, the British met little opposition in their land grab.

The word *variola* is derived from the Latin *varius*, meaning speckled. To variolate is to inoculate with smallpox.[5] The word was first introduced in AD 750 by the Swiss Bishop Marius of Avenches, near Lausanne, Switzerland. However, centuries passed before Europe adopted variolation. Aside from prevention through quarantine, it was the only effective means of averting a major infection.

By the time of the American Civil War, smallpox vaccination was widely practiced. But because the vaccine did not give lifelong protection, it had to be re-administered every 10 to 20 years in order to prevent an epidemic. However, the families on the lower end of the socio-economic strata found the cost of vaccination prohibitive and often suffered the consequences. Additionally, the low incidence of smallpox in the United States by the 1840s caused many people to forego vaccination. Though it was a requirement for Union and Confederate troops, this regulation was not always practiced, and smallpox outbreaks were reported, especially in prison camps. Between May 1861 and June 1866, nearly 19,000 smallpox cases plagued both armies and claimed the lives of about 30 percent of its victims. This led to arm-to-arm vaccination among the troops, in which matter from smallpox pustules was delivered from one soldier directly to the arm of another, but this had all of the drawbacks of blood barrier contamination. In one case, 5,000 Confederate troops were infected with syphilis as the result of the arm-to-arm technique, taking them out of duty. For this reason, children were often used, as they carried fewer blood-borne diseases.

««« »»»

As early as AD 1000, inoculation against smallpox was common in China and India. It was well known among medical practitioners that those who survived the dreaded disease developed immunity to it, so the Chinese ground the dried scabs from smallpox pustules into a fine powder and inhaled them through a straw. Patients developed a milder form of smallpox and benefitted from lifelong immunity to the disease. In a small percentage of cases, however, the inoculation developed into full-blown smallpox, resulting in death. Smallpox in its milder form still carried the side effect of disfiguring pockmarks. Nevertheless, it was an improvement over suffering the more virulent form of the disease.

The 15th century brought a new method of variolation to the Middle East: powdered smallpox crusts were inserted into a small incision with a pin. The patient developed pustules around the incision and a mild form of the disease, but avoided the disfiguring consequence of inhaling the powdered scabs. Approximately 1 in 100 patients died from the procedure.

In 1721, Lady Mary Wortley Montague, the wife of the British Ambassador to Turkey, brought the Turkish method of variolation to England. Considered a beautiful maiden in her youth, her face was hideously blemished by the disease at age 26. Her son is considered the first European to have successfully undergone variolation against smallpox. Delighted by the technique, she campaigned to bring inoculation to England. To prove its effectiveness, her 4-year-old daughter was inoculated in the presence of physicians from the royal court. They were so impressed by her mild case of smallpox and subsequent immunity to the disease that they made it a common practice among British royalty.

Although the royal court embraced variolation as a method of preserving heirs to the throne, the medical establishment resisted the change. Even more steadfast was the clergy's insistence that variolation interfered with God's will. However, the British royalty affronted the religious leaders by having all of its soldiers inoculated.

When smallpox killed members of continental Europe's royal families, acceptance of variolation became widespread among society's upper crust. Empress Marie Therese of Austria protected her extended family through variolation. Louis XVI of France and his children were inoculated. Catherine II of Russia and her son followed suit. And Frederick II of Prussia ordered variolation for all of his troops.[6]

Inoculation hospitals offered variolation primarily to the affluent, because they could afford the lengthy procedure. It included a preparatory period of fasting, purging, and bloodletting, followed by the variolation procedure. Eight or nine days later, the patient came down with a usually mild case of smallpox, followed by a prolonged period of recuperation. However, it was a risky procedure that transmitted full-blown symptoms to some of its weaker patients, delivering a fatal blow to a small percentage of them. Recently treated patients could easily infect others and therefore had to remain confined in inoculation hospitals as long as their skin rash persisted. The technique was somewhat successful, but human diseases such as syphilis crossed the blood barrier.

The working class could not afford variolation and suffered from the epidemic disease at an alarming rate. The mortality rate was extraordinarily high among young smallpox victims, killing hundreds of thousands each year.

Unfortunately, variolation did not extend to the crew aboard cargo ships. A British merchant ship, sailing from Salt Tortuga in the Caribbean, arrived in Boston, Massachusetts, in mid-April 1721, with more than its expected cargo. As was the custom, the crew underwent a quarantine inspection for smallpox, passing with flying colors. The sailors were free to make their way about town. But within a day, one crew member developed acute symptoms of smallpox, and by early May nine more were ill. Although they were confined to a quarantine house by the dock, the disease had already spread to Boston's residents. Soon thereafter, hundreds fell ill with the epidemic that raged until the following spring.

Boston's Reverend Cotton Mather, vicar of the Second Church of Boston and lay physician, had learned about variolation from his

African slave and convinced his friend, Dr. Zabdiel Boylston, of its effectiveness. Boylston's first patient, his own son, recovered from the mild form of the disease and remained immune to it when exposed to the active virus. Convinced, Boylston performed variolation on 247 patients. Only six, who had weakened immune systems, developed the full-blown form of the disease, resulting in their deaths.[7] That same year, 1721, the smallpox epidemic afflicted nearly half of Boston's population of 12,000 and claimed the lives of 884 victims.

Boston's medical establishment, led by Dr. William Douglass, a staunch opponent of variolation, engaged in a bitter struggle against smallpox inoculation. Convinced by the negative propaganda espoused by Douglass and his cohorts, the press engaged in a campaign against variolation on the grounds that it interfered with divine providence. The Anti-Variolation League demanded the death penalty for physicians who dared practice inoculation. Dr. Boylston's life was threatened and he could not safely leave his home in the evening. A fire bomb was hurled into Reverend Mather's home. Opposing clergy issued the statements, "...for a man to infect a family in the morning with smallpox and to pray to God in the evening against the disease is blasphemy: [that the smallpox is] a judgment of God on the sins of the people, [and that] to avert it is but to provoke him more."[8]

However, many clergy members were strong supporters of Dr. Boylston's efforts and underwent variolation themselves. Soon it became apparent that variolation was superior by far to contracting the more virulent form of the smallpox virus, and Dr. Boylston achieved notable success. He pub-

Vaccination is derived from the Latin word *vaccinus* meaning "of cows"[9] because the smallpox vaccine is derived from cowpox pustules.

lished his results and was elected to the British Royal Society in 1726. Back in the colonies, however, he continued to face limited opposition for another 20 years. His most tenacious opponent, William Douglass, eventually accepted variolation, but never retracted his malicious remarks about Boylston.

In 1774, Benjamin Jesty, a British farmer who survived smallpox as a child, became concerned for his family during an outbreak of the disease. He was aware of the fact that people who had contracted cowpox, a mild disease in comparison, developed immunity to smallpox. His two milkmaids had both suffered cowpox lesions on their hands, and cared for their family members during the smallpox epidemic without ill effect. Jesty reasoned that if he deliberately infected his own family with cowpox, he could save them from the dreaded and often fatal disease. A fortuitous outbreak of cowpox in a neighbor's herd brought Jesty and his family to the pasture, where he scraped the pustules from the cow's udder and inserted the pus and fluid into his family's arms with the aid of a stocking needle. His two sons developed a mild form of cowpox, but his wife was not so fortunate. She developed a raging fever and her arm became severely inflamed. Luckily, she made a full recovery and lived for another 50 years.

As is so often the case, what could have been a success story was met with scorn by the Jesty's superstitious neighbors. As word of the inoculations traveled throughout the village, Jesty became the target of contempt and harassment, enduring hoots and pelting in the marketplace. Jesty later moved his family to the Isle of Purbeck, where he successfully performed cowpox inoculations on many of the island's residents. However, he would not go down in history as the father of smallpox vaccination. That honor was awarded to a man who adhered to the tenets of science—Dr. Edward Jenner.

Edward Jenner was the son of Stephen Jenner, an Anglican minister in Berkeley, Gloucestershire. The eighth of nine children, he was orphaned by age 5 and went into the care of his oldest sister, Mary, and her husband, Reverend G.C. Black. In 1764, Jenner began an apprenticeship with Dr. George Harwicke. In 1770, at age 21, he trained in London under John Hunter at St. George's Hospital. Hunter was considered a great surgeon and experimentalist. Under his guidance, Jenner began experimentation using scientific methodology. In 1772, Jenner returned to Berkeley and began his practice in general medicine and surgery. A man of many interests, he entered into additional scholarly pursuits in the natural sciences. He was also an outstanding musician and poet. Upon his return to his native home in 1772, he surveyed the geography of his county and studied the migration patterns of birds. Additionally, he was one of the first in his area to construct and test a hydrogen balloon. His 1788 paper, "Observations of the National History of the Cuckoo," earned his election to the British Royal Society.

Edward Jenner, 1749–1823. Courtesy of the National Library of Medicine.

Jenner sought a safer form of protection from smallpox than variolation. As a child he endured a lengthy and traumatic variolation treatment which had caused him considerable suffering. Haunted by the severity of his own infection and the several months of recovery time, he sought methods to either cure or prevent the dreaded disease. He began by collecting anecdotal evidence from patients and anyone else who could contribute relevant information. As a young apprentice he had heard about the folk remedy used by Jesty and others. In 1770, he was intrigued when a milkmaid advised him that cowpox, a milder infection, had protected her from smallpox. However, his desire to conduct experiments was delayed by the absence of cowpox cases in the area, until the cowpox epidemic of 1796. In an experiment that would today be considered unethical, Jenner obtained permission to inoculate James Phipps, the 8-year-old son of his farmhand, with matter from cowpox pustules. On May 14, 1796, Jenner made two half-inch incisions in young James's arm and introduced fluid from a cowpox lesion on the hand of Sarah Nelmes, a young milkmaid.[10] The affected area around the site of the inoculation became swollen and blistered, but healed within two weeks, leaving only a small scar. In the second half of his experiment, on July 1, Dr. Jenner inoculated young James with live pus and fluid from an active case of smallpox. The boy exhibited no symptoms of the disease.

In 1797, Jenner submitted a paper on his findings to the British Royal Society, which in 1788 had published his work as a naturalist. His outstanding work in natural history had secured his induction into the elite British Royal Society, but his paper on smallpox was rejected on the grounds that his ideas were too revolutionary and his experiments too limited. The society's leaders advised him not to publicly disseminate his credulous ideas if he valued his reputation. Convinced that his scientific evidence held merit, in 1798 Jenner self-published a 75-page pamphlet, "An Inquiry into the Causes and Effects of the Variolae Vaccinae, A Disease Discovered In Some Of The Western Counties of England, Particularly Gloucestershire, And Known By The Name Of Cowpox." It described his groundbreaking work with 23 patients—10 of whom he inoculated with fluid from cowpox pustules and 13 who developed subsequent immunity to smallpox after suffering from cowpox. Although he aroused some positive interest, his discovery stirred raging controversy among the clergy and the medical establishment.

Reverend Thomas Robert Malthus, the father of the science of political economics, rejected the notion of vaccination. Proclaiming that smallpox was nature's remedy for overpopulation and God's method of reducing the numbers of poor people, he vehemently rejected vaccination. His 1798 "Principle of Population" essay warns that passion between the sexes is a necessary evil that will continue to result in overpopulation, especially among those who do not exercise restraint. Furthermore, unimpeded population growth will lead to famine. Only misery and vice, in his opinion, provided the checks and balances necessary to keep the world's populations from outgrowing obtainable sustenance. He proposed the idea that infant mortality, wars, plagues, famines, and disease were the only solution to impending starvation.

The popular media had a field day with Jenner's vaccine, calling the introduction of bovine material into humans an abomination. A print produced by political satirist James Gillray depicts Jenner in a room full of patients in various stages of growth, "sprouting" cow parts. Jenner is depicted vaccinating an apprehensive woman in the foreground. To her right, a man holding a pitchfork has calves emerging from his buttocks and arm. The pregnant, (or could it be an udder?) woman behind him is choking on the calf ejecting from her mouth. The man behind her has grown horns and a bovine face. This common notion of becoming part cow, promoted by the media through cartoons and songs, led to a widespread hysterical reaction among uneducated segments of society. The associated fear factor dissuaded many from obtaining vaccinations and led to smallpox epidemics and innumerable deaths.

"The Cow Pock-or-the Wonderful Effects of the New Inoculation! Vide-the publication of ye Anti-Vaccine Society," by James Gillray. Courtesy of the National Library of Medicine.

Dr. William Woodville was the first physician to carry out a large-scale vaccination experiment at his Smallpox and Inoculation Hospital in London. It was not successful. Of the 600 people he vaccinated in early 1799, 400 developed the multiple lesions suffered in the variolation process. Jenner attributed the outbreak to unsterilized lancets contaminated with smallpox. It was only after Dr. Woodville noted that none of the patients vaccinated in his private office developed a mild case of smallpox that he realized Jenner was correct. He then supported Jenner's discovery.

As you can see, fear mongers, many of whom had a financial stake in variolation hospitals, attempted to suppress Jenner's discovery. But others were impressed by his careful observation and documentation. England's royal family quickly replaced variolation with the new procedure, vaccination, as did many physicians throughout Great Britain. In 1802, the British Parliament awarded Jenner what would today amount to $500,000. Five years later, the it granted him twice that amount. He was honored by Oxford, Cambridge, and Harvard and inducted into several scientific societies as an honorary member. Jenner hoped only to alleviate human suffering brought on by the most lethal scourge know to mankind. His brilliant scientific experimentation and dogged perseverance led to the eventual total eradication of smallpox nearly two centuries later.

Jenner's scientific methodology so impressed Dr. Benjamin Waterhouse, a professor at Harvard Medical School, that Waterhouse obtained cowpox fluid from England and vaccinated his family and servants. Then he exposed one of his servant boys to smallpox at a local quarantine hospital. Jenner's observation was confirmed when the servant remained free of the disease.

Waterhouse published a pamphlet on his experiments, *On the Prospect of Exterminating the Smallpox*. But his effort was met by repudiation from the American medical establishment, who had a financial interest in variolation. However, experimental trials clearly demonstrated the superiority of vaccination over variolation, and Waterhouse prevailed. Soon doctors clamored for his vaccine, but he selectively distributed it and expected to share in the profits.

President Thomas Jefferson was so impressed by Waterhouse's pamphlet that he had 18 members of his family, his servants, and some neighbors vaccinated. When Chief Little Turtle, chief of the Miami Tribe, and his warriors visited the White House, Jefferson offered to vaccinate them against smallpox. They accepted. He appointed Dr. Waterhouse Vaccine Agent in the National Vaccine Institute to establish vaccination throughout the United States.

The king of Spain, Charles IV, considered it so important to the preservation of the Spanish economy that in 1803 he developed an official vaccination program. Residents in Spanish colonies throughout the Americas were vaccinated against smallpox. Napoleon followed suit, first by vaccinating his army and then French civilians. The policy spread to Bavaria and Denmark, and finally throughout Europe and Russia.

The greatest difficulty was in obtaining and preserving the live vaccine. Cowpox was rare, and when the pustule fluid was obtained, it was rapidly inactivated by exposure to sunlight or high temperatures. This problem was overcome with arm-to-arm vaccination. As mentioned previously, that presented a new problem when adult donors passed diseases such syphilis through arm-to-arm vaccination. This problem was eliminated when it was discovered that the cowpox virus could be inoculated directly into the skin of a calf through a surgical lacerations. This process made it possible to safely harvest large amounts of vaccine directly from cows.

As nations began to institute mandatory vaccination policies, anti-vaccination societies sprang up throughout Europe. Opponents questioned its safety and proclaimed it a violation of civil rights. As late as 1889, Edward M. Crookshank, professor of Comparative Physiology and Bacteriology at Kings College, London, wrote, "[Cowpox does not] exercise any specific protective power against human Small Pox."[11] The same year, Charles Creighton, a British medical historian, wrote, "[Cowpox] was fancifully represented as an amulet or charm against smallpox. By the idle gossip of incredulous persons who listened only to the jingle of names... Let us suppose that the glowing end of a cigar be firmly applied to an infant's arm;...[a]sore will result which may be called cigar pox... [T]he cigar pox is in its pathology just as relevant to the smallpox as the cowpox is."[12] This opposition led to the European smallpox epidemic of 1870–75, killing thousands.

The age-old "spontaneous generation of disease" theory was abruptly altered with the acceptance of French scientist Louis Pasteur's germ theory of disease. Early in his career, he discovered that microorganisms cannot be spontaneously generated in nutrient broth without exposure to outside spores and dust. Therefore, it is microscopic organisms—not miasmas—that cause disease. But his discovery was not immediately accepted. Many influential opponents from the scientific establishment clung to their archaic belief that disease was the result of a poisonous atmosphere rising from swamps and putrid matter. They were not quick to acknowledge that his germ theory of disease was valid. It was Robert Koch's series of proofs that anthrax was caused by a bacterium, and Pasteur's subsequent development of a vaccine against it, that finally verified germ theory. This was further supported by Pasteur's successful demonstration that his fowl cholera vaccine was effective. In 1882, Pasteur was elected to the Academie Française. The Louis Pasteur Research Institute opened in 1888, which he headed until his death in 1895.

The mid-20th century brought advances in maintaining potent vaccine through refrigeration and freeze-drying, enabling more people to be vaccinated. In the 1950s, Europe and North America launched a concerted effort to eliminate smallpox through mandatory vaccination. Although the campaign wasn't 100-percent effective, it made great strides in eradicating the disease in the developed world. However, smallpox epidemics raged unabated in less developed tropical and sub-tropical countries, primarily as a result of vaccine instability, organizational difficulties, and underfunding. In 1958, The World Health Assembly received an alarming report about the devastating impact of smallpox in 63 countries. In response, the Soviet Union proposed a program for the global eradication of the disease. An attempt was made to eliminate smallpox worldwide, but it wasn't until the World Health Organization began a global campaign in 1967 that significant progress was made. The additional funds funneled into the organization intensified the worldwide effort, and smallpox was finally wiped off the face of the earth. The world's last natural case of smallpox was reported in Merka, Somalia, in 1977. A year later, a medical photographer at the University of Birmingham died after she was accidently contaminated by smallpox. Her mother also developed the disease, but survived. On May 8, 1980, The World Health Organization announced that smallpox had been obliterated from the face of the earth. Edward Jenner's perseverance in the face of adversity triumphed in the eradication of the world's most lethal plague.

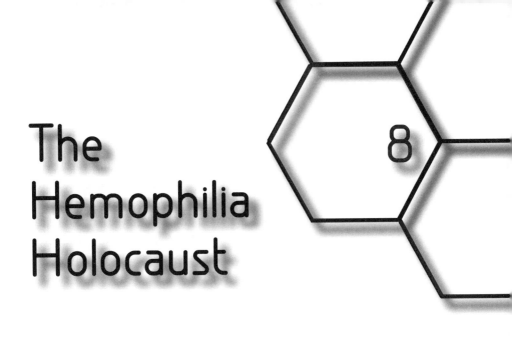

The Hemophilia Holocaust

8

In 1980 gay liberation was in full swing on America's liberal coast, but homophobia, burgeoned by religious fundamentalists, was sweeping the heartland. Ronald Reagan had defeated incumbent Jimmy Carter in a landslide victory, and Jerry Falwell's Moral Majority was taking credit for the triumph. It was not only a political conquest, but the voice of a pro-family, anti-gay agenda. The more liberal segments of society were suggesting that the incumbent's failures had led to his defeat, not a moral mandate. They pointed to the success of liberal homosexual enclaves across America in gaining political power and reducing homophobia. During this time, young Adam Sanderson, a hemophiliac, watched with interest as his older brother, Jeffrey, marched for gay liberation in the streets of San Francisco. No one could foretell that both would become the victims of a deadly viral holocaust.[1]

Adam inherited hemophilia, a genetic disorder that has been noted as far back as biblical times. His maternal grandfather had inherited the disorder through his mother's defective chromosome, which was carried only on one of her XX chromosomes. The defective chromosome was then passed to Adam's mother. She in turn had the potential of passing the defective chromosome to half of her son's XY. Adam acquired the defect, but Jeffrey did not. Their sister had a 50 percent chance of inheriting the defective gene, but not the bleeding disorder. She, in turn, carries the risk of passing hemophilia onto her sons.

Hemophilia affects a very small percentage of the population—approximately only 20,000 in the United States. There are two main types of hemophilia: A and B. Approximately one in 5,000 male babies in the United States is born with hemophilia A, whereas hemophilia B occurs in only one out of 33,300 male births in the United States. It is caused by a deficiency in clotting Factor IX. Adam has hemophilia A and therefore is deficient in clotting Factor VIII.

Hemophilia is not consistent throughout the affected population and can range from mild to severe forms. People with mild hemophilia usually have problems only after a serious injury or a surgical procedure and may not be diagnosed with the disorder until adulthood. Moderate hemophilia causes bleeding after injuries and can also trigger spontaneous bleeding episodes without an apparent cause. Severe hemophilia causes bleeding after injuries and frequent spontaneous bleeding episodes, often into the joints and muscles. Adam has the most severe form. This means that he can experience very painful, uncontrolled bleeding into his joints, such as knees, elbows, and ankles, from a minor bruise, bump, or strain. As a result, his joints become very painful and swollen. After repeated episodes of bleeding into the joints, Adam could lose joint mobility. Ice can be useful to relieve the pain and swelling, but blood transfusions are the standard course of treatment.

There was a major breakthrough in the treatment of hemophilia when Adam was a youngster. Dr. Judith Poole, a post-graduate student at Stanford, noticed unusual crystals forming on the top of plasma she was thawing. She tested them and found that they were made up of pure fibrinogen, a protein essential for coagulation. Named *cryoprecipitate Factor VIII concentrate*, it was 30 times more effective in blood coagulation than fresh frozen plasma. Because the process was done in small batches with only a few donors in local blood banks, the possibility of contamination by viruses was low.

Soon, however, drug companies developed sophisticated chemical processing techniques to extract the Factor VIII from large batches—up to 20,000 units of blood—and produced a powder that had a high concentration of Factor VIII. A couple of little bottles contained hundreds of units of Factor VIII, and they could determine the actual number of concentrate units per bottle, which helped to resolve the question of how much to give to each patient. This new Factor VIII concentrate controlled the bleed by bringing Adam's blood factor level up to 50 percent of the normal range, which usually resulted in coagulation. The freeze-dried fibrinogen powder needed only to be combined with distilled water and administered intravenously in hospitals.

A major breakthrough came when a home treatment program that could be administered by trained family members was approved. Adam's parents learned to transfuse their son when he was young and he learned to transfuse himself in his early teens. Although the product was quite expensive, costing thousands of dollars per year, it was far superior to the earlier method of hospital transfusions of whole blood or fresh frozen plasma. Many hemophiliacs came to lead almost normal lives with far less pain than they experienced in the pre–Factor VIII concentrate time.

Michael Sanderson, Adam's father, had been a vocal advocate for safety in the blood supply and personally researched prophylactic measures in development all over the world. He sat on local medical boards and often supplied board members, including doctors, with the latest research that had been published in medical journals. A new drug discovered in Japan was showing some promise. The chemical name is epsilon amino-caproic acid (EACA), a synthetic chemical agent that completely inhibits a molecule responsible for the degradation of fibrin (clot dissolving). It is generally taken orally and is a powerful antifibrinolytic (stops the body from dissolving its clots) agent. Two doctors from Philadelphia had done controlled studies with EACA (trade name Amicar, as made by A.B. Kabi of Sweden), when their hemophilia patients experienced joint bleeds.

Everybody close to hemophiliacs, including parents, doctors, other family members, and the hemophiliacs themselves, was well aware that there seem to be more frequent painful bleeds when the patient is under stress. Furthermore, two episodes of the same trauma, such as a bang on the elbow, will sometimes produce a bleed and sometimes not. So, absence of sufficient clotting factor is a necessary but not sufficient condition to produce a bleeding episode. The doctors who spearheaded this EACA research found that the effective half-life of the drug was just a few hours. Then the poorly formed clot would dissolve and the hemophiliac would be back in trouble. Because fibrinolytic (clot dissolving) activity varies considerably, the dosage had to be measured to suit the patient's needs. It was found that the drug, which was available as a bad-tasting liquid or in pill form, had to be taken for at least a couple of days at four-hour intervals. But it worked.

Dr. Louis K. Diamond is referred to as the father of pediatric hematology for establishing one of the world's first pediatric hematology research laboratories at Children's Hospital in Boston. He was a professor of pediatrics of the Harvard Medical School, and later, at the University of California. Having directed the Red Cross's National Blood Program from 1948–1950, he received several awards for his work on blood banking and safe blood transfusions.

Michael Sanderson called the doctor who published the paper, and then called his son's medical doctor, who wrote a prescription for it. Although the drug was not commonly used as a prophylactic measure to treat hemophiliacs, Adam's father considered it less risky than the frequent use of Factor VIII concentrate. They used it episodically whenever Adam experienced a bleed rather than on a regular basis. Despite its bad taste, Adam and his parents preferred the liquid Amicar, primarily because attempting to administer eight pills to a child in the middle of the night was not a pleasant chore.

Prior to the development of Amicar, when Adam sliced his finger on a potato peeler he had to be transfused with Factor VIII concentrate every four hours for two days. A tongue bite or joint bleed sent him to the hospital because he required around-the-clock transfusions. With Amicar, Adam produced only one drop of blood after a tongue bite. He didn't require a hospital transfusion for five full years. It showed great promise.

When Adam was successfully using EACA, Dr. Louis K. Diamond published a paper titled "Ineffectiveness of EACA in the Treatment of

Hemophilia" in the prestigious *New England Journal of Medicine.* An editorial in the same journal delivered the disappointing message that a hoped-for solution to hemophilia had once again failed, but the study that it spoke of was seriously flawed, in Michael Sanderson's opinion.

In Dr. Diamond's clinical trials, patients were given much lower doses than should have been used. They were not administered the drug in four-hour intervals during the night, and the drug was stopped when there was an obvious bleed. Michael Sanderson reasoned that most people would not find much benefit in the use of aspirin to cure a headache if half a dose was administered every 24 hours. The study results represented a major setback in Adam's treatment and a huge disappointment to the family who had been using EACA successfully.

Unfortunately, the North American medical community ignored the effectiveness of EACA as an adjunct treatment for hemophilia. It could not move beyond the idea that hemophilia is a disorder characterized only be a reduced capability to make the Factor VIII. So their only solution was to replace the Factor VIII—the greater the severity of the hemophilia, the more Factor VIII the patient required.

Once again, Adam had to rely exclusively upon Factor VIII concentrate to control his bleeding episodes. Because he was a severe hemophiliac, he required multiple transfusions with Factor VIII concentrate each month. Thus, he couldn't avoid being exposed to viral contamination from thousands of blood donors. Alarm swept through the Sanderson household in 1981 when the Centers for Disease Control in Atlanta were alerted to a new disease affecting a sector of America's blood-donor community.

A small percentage of the population had come down with perplexing symptoms, including red, purple, brown, and black raised lesions on the skin and in the mouths of mostly gay and bisexual men. Diagnosed as Kaposi's sarcoma, caused by the Human Herpes Virus 8, the malignant lesions were spreading to organs in the body such as the lungs, liver, and intestinal tract. Prior to this outbreak, Kaposi's sarcoma had been considered a benign cancer, prevalent primarily in elderly Jewish and Italian men, and sometimes in Africa. In most cases, the elderly men had died from other causes many years later. But something new and unprecedented was happening. Seemingly healthy men were beginning to waste away and die at an explosively rapid rate.

In November 1980, a 31-year-old gay patient, whose immune system appeared to be almost nonexistent, was diagnosed with Pneumocystis carinii pneumonia, a rare lung infection caused by a protozoan organism common in the environment. The symptoms included rapid breathing, fever, shortness of breath, and dry cough, and it was beginning to emerge in the same gay and bisexual group that was at risk for Kaposi's sarcoma. Suddenly and without explanation gay and bisexual American men were dying from the disease.

Most perplexing was the fact that, although both diseases were usually limited to those whose immune systems were compromised, seemingly robust men were beginning to drop like flies. Scientists began to suspect that a new disease was attacking the immune systems of this specific population. Although the virus had not yet been isolated, it appeared that the disease might be transferred via body fluids, which included semen, urine, and saliva. It had been suggested that this disease might also be passed in blood products.

A Florida physician issued an alert to Dr. Bruce Evatt, the resident expert in hemophilia at the Center for Disease Control, that his heterosexual hemophilia patient had died from Pneumocystis carinii pneumonia. He speculated that if blood products were contaminated with Pneumocystis protozoa, Factor VIII concentrate could infect hemophilia patients with the new, life-threatening disease. But Dr. Evatt reasoned that Pneumocystis protozoa were too large to have passed through the filters in the preparation of Factor VIII concentrate. This was the same filtering process that cleaned bacteria out of the blood but allowed smaller microbes such as the hepatitis B virus to remain in the clotting factor.[2] There was no compelling evidence that a virulent retrovirus capable of eliminating one's immune system was lurking in the blood supply. Therefore, he didn't think that the hemophilia population was at risk.

The emerging pandemic that had acquired the acronym GRID (later changed to AIDS), standing for "gay-related immune deficiency," was beginning to afflict gay enclaves along the east and west coasts of the

United States. Europe and Africa had also seen a smattering of cases in both heterosexual and homosexual populations. The mortality rate was near 100 percent for those unfortunate enough to become symptomatic. Immediate funding on a large scale would be required to identify and eradicate the cause of GRID.

But GRID research was in gridlock. Virologists at the Center for Infectious Diseases at the Centers for Disease Control thought that a retrovirus explanation was highly unlikely.[3] Dr. Don Francis, a leading epidemiologist, speculated that he was observing feline leukemia, only in humans. The virus destroyed the immune systems of cats and left them open to infection and a variety of cancers. He wanted to put blood banks on alert because he thought that it would most surely show up in blood. But he had no proof.

To make matters worse, funding was in short supply. The anti-gay Reagan administration did little to fund research for "gay cancer" and quietly promoted the idea that God was punishing homosexuals for their lifestyle choices. Blood banks were resistant to the idea that their blood products might be contaminated, and professional medical journals refused to publish articles on the disease without proof that it was infectious.

In July 1982, GRID was renamed Acquired Immune Deficiency Syndrome to eliminate the social stigma associated with the "gay" disease, and because it was increasingly apparent that it was widespread in the population. AIDS was a slowly developing disease, which meant that a single infected person, through sexual activity, could transmit the virus to many people without knowing he or she was a carrier. The incubation period could range from two to seven years before the virus struck with a vengeance. In the interim, an asymptomatic carrier could innocently infect all of his or her sexual partners. The insidious disease was beginning to spread rapidly throughout the gay, IV-drug user, and Haitian communities in the United States, and among women who had sex with the high-risk groups. It would soon kill their babies as well. The news was particularly distressing to the Sandersons. They had two reasons to worry: Adam's Factor VIII concentrate might be contaminated, and Jeffrey was openly gay.

By the end of 1982, the Centers for Disease Control reported the first cases of AIDS in the hemophilia population, probably transmitted through Factor VIII concentrate. Yet the blood banks were fiercely resistant to the idea of contacting or questioning their donors, and they challenged the CDC to prove that AIDS was transmissible through blood products. Everyone required proof beyond the shadow of a doubt that AIDS was blood borne in order to take steps to protect the recipients of blood products. No one wanted to alarm the general public, or perhaps more importantly, to risk revenue losses.

Additionally, officials at the blood banks and pharmaceutical companies continued to reassure hemophilia patients their Factor VIII clotting substance was safe. They reported that the chance of contracting AIDS from their products was only one in a million. One of the reasons for the advice to the hemophilic community to continue as usual was that the doctors didn't know what else to do if transfusions were stopped. They also didn't seem to understand that if any of the blood units were contaminated with viruses such as hepatitis or HIV, the entire batch would be contaminated because the viruses thrive in the blood.

For the Sandersons, 1982 was an unfortunate year. Jeff had recently fallen ill with a lung infection and had to be hospitalized. The diagnosis was Pneumocystis carinii pneumonia. He didn't last long. In October 1982, the Sanderson's buried their eldest son.

Michael and Judy Sanderson were well aware of the possibility that Adam could become infected as well. But their hands were tied: there was no viable medical alternative for hemophiliacs.

In October 1982, pediatric immunologist Dr. Art Ammann spoke at the first national conference on AIDS in San Francisco, California, about an Rh-negative baby born on March 3, 1981, who had been transfused over a seven-day period with blood from 13 donors. A week later, special cells that aid in blood clotting were transfused into the baby at the University of California Medical Center.[4] By September, the 7-month-old baby was suffering from a perplexing immune dysfunction that wracked his tiny body with one opportunistic infection after another. A month later, he was diagnosed with AIDS. In November, the blood bank responsible for the baby's transfused material completed its blood donor records search: one donor had died from AIDS. Given the fact that this baby could not have engaged in IV drug use or sexual activity, the proof that he had acquired AIDS through a blood transfusion seemed incontrovertible. This finding commanded action.

When CDC epidemiologist Dr. Don Francis sat down with representatives from the major pharmaceutical companies and blood banks on January 4, 1983, he hoped to convince them to take steps to protect the blood supply. But the billion-dollar-a-year blood industry vocally rejected the evidence that AIDS was blood borne and rationalized their decision to conduct business as usual. An outraged Dr. Francis delivered an emotional plea, "How many people have to die? How many deaths do we need? Give me a threshold of death that you need in order to believe that this is happening, and we'll meet at that time and we can start to do something." He would later describe the meeting as a tacit endorsement of negligent homicide. [5]

The following day, an editorial in the *New England Journal of Medicine* recommended a radical change in the treatment of hemophiliacs and in the U.S. National Hemophilia Foundation. Promoting the assumption that a high percentage of Factor VIII concentrate was contaminated with the AIDS virus, the writer issued recommendations for the prevention of AIDS in hemophiliacs. This seems to have pressured some blood bankers into screening out high-risk groups, but they incorrectly assumed the low-risk category was not infected. Simultaneously, vocal members of the high-risk groups were crying discrimination. Civil rights issues and the blood industry's profit motive trumped common sense and the common good.

This changed in March 1983, when all private pharmaceutical companies instituted a screening process to eliminate the high-risk groups from their blood supply. That same month, the Department of Health and Human Services in Washington, D.C., issued a temporary measure under the auspices of the Center for Disease Control, the National Institute of Health, and the Food and Drug Administration. It stated that sexually active gays with overt symptoms of AIDS, or those who had engaged in sexual relations with those who did, should not give blood. The blood bankers were left with no other alternative except to comply with the new directive. Unfortunately, in the first three months of 1983, nearly a million blood transfusions, many contaminated with AIDS, had been administered in the United States.[6]

As a result of proclamations that the chance of the virus being transferred by the concentrate was only one in a million, and that no drug was safe enough to take the place of a transfusion, hemophiliacs in the United States and Canada were at great risk for contracting AIDS. It was during this period of denial that young Adam, along with 10,000 hemophiliacs, were transfused with AIDS-contaminated Factor VIII concentrate.

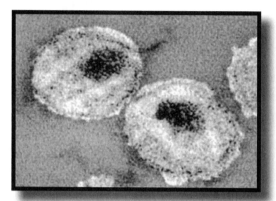

The HIV Virus. Courtesy of the National Institutes of Health.

It turns out that, contrary to the well-meaning but deadly advice of the doctors advising the hemophilia community, there were alternatives to AIDS-contaminated Factor VIII concentrate. Some European pharmaceutical companies were routinely heat-treating it to kill the hepatitis B virus. They would later find out that it also killed the AIDS virus. A German company had developed and was marketing the first pasteurized Factor VIII. France, which was on the cutting edge of AIDS research, had banned the importation of American blood products due to the risk of contamination. The Dutch and British were considering the same move. But American pharmaceutical companies were concerned about the high cost of heat-treating plasma and the up to 25-percent product loss during the heat-treating process. It simply wasn't cost effective. After considerable time, heat treatment was used, but thousands had already been infected with the deadly virus.

At the Stanford University Blood Bank, Dr. Edgar Engelman was beginning to think that the "one in a million" chance of contracting AIDS from a blood transfusion was a cruel hoax.[7] He realized that the blood banks were manipulating data to suit their own financial interests by grossly underestimating the number transfusion-related AIDS victims. They were simply ignoring the pre-AIDS cases and counting only those with the full-blown disease. He estimated that the chance of contracting AIDS from a San Francisco blood transfusion was conservatively one in 10,000, and perhaps one in 5,000.[8]

Meanwhile, two French researchers, Dr. Luc Montagnier and Dr. Françoise Barre at the Louis Pasteur Institute in Paris, France, had made great strides in their attempt to identify a retrovirus as the cause of AIDS. They thought that they might be on to something, but each time they ran the radioactive test to detect the presence of the chemical that retroviruses secrete in order to reproduce, the count was too low.

Finally, by mid-January 1983, the radioactive measure came high enough to positively identify Barre's sample as a new retrovirus, but more tests had to be done to validate the discovery. The French sent off samples to Robert Gallo's laboratory at the National Cancer Institute. Luc Montagnier had already submitted a professional research article to *Nature*, but he was not successful in getting it published. Gallo agreed to submit the Pasteur team's research findings to *Science* magazine, along with his own.

The French had already developed a test to detect antibodies to the AIDS virus in 1983, but political posturing stood in the way. The tragedy is that the blood supply continued to be infected, and thousands of people died as a result of delays.

By the end of February 1984, the samples and information supplied by the Louis Pasteur Institute to the Centers for Disease Control in Atlanta were used by the National Cancer Institute to make antibody tests for the virus. Robert Gallo was incredulous about the French scientist's discovery and even suggested that their samples had been contaminated with his virus. Gallo soon announced that he had independently discovered HTLV-III, the true cause of AIDS, and that his discovery differed from the French virus.[9]

On April 23, 1984, Margaret Heckler from the U.S. Department of Health and Human Services announced that Robert Gallo from the NCI had isolated the virus that caused AIDS through a collaborative effort between the Louis Pasteur Institute and the National Cancer Institute. It was a false and misleading statement that led to an international scientific crisis. In fact, the discovery attributed to Robert Gallo had come a full year after the Louis Pasteur Institute announced they had discovered the virus.

Less than a year later, the public and the scientific world received the news that Gallo's sample was genetically identical to the French sample. It could not have been donated by anyone but the French institute's patient. It appeared that Gallo had used virus specimens and research data supplied by the Louis Pasteur Institute's research team to isolate the AIDS virus and develop the viral antibody test. Gallo was outraged by the suggestion of wrongdoing and countered that it had been impossible for his assistants to grow the French viral cultures.

Although the French researchers applied for an AIDS blood-test patent in December 1983, five months before Gallo submitted his, Gallo was named chief holder of the patent. Royalties from the sale of the test went to Gallo and the U.S. Federal Government. The Louis Pasteur Institute filed a lawsuit against the National Cancer Institute and the U.S. government in December 1985 for breach of contract, alleging that Gallo had violated a promise to use French samples for research purposes only. Facing the possibility of an international embarrassment, the Reagan administration intervened and reached a compromise with French President Jacques Chirac in March 1987. Robert Gallo and Luc Montagnier were named as co-discoverers of the newly named HIV virus and the French shared the royalties.

Only eight days later, the Los Alamos National Laboratory in New Mexico sent a confidential memo to the National Institutes of Health (NIH) informing them that Gallo's virus samples could not have been isolated from U.S.-based blood samples. The samples clearly belonged to the French.[10] However, the report was promptly buried at the NIH and remained there until it was accidentally discovered in 1994.

In 1991, the National Institute of Health commissioned a series of experiments to once and for all determine which party had discovered the HIV virus. The test results clearly supported the Louis Pasteur Institute's claim that they had been the first. None of Gallo's patients could have supplied the genetic material that he claimed was cultured in his lab.

After years of legal battles, often obfuscated with false statements and redacted documents, the National Institute of Health agreed to award the Louis Pasteur Institute the major percentage of future royalties for their part in developing the AIDS test. It was already 1994, and years of legal wrangling had obstructed the serious business of making strides in AIDS research.

In a final attribution of credit, the Nobel committee awarded Françoise Barre-Sinoussi and Luc Montagnier, the two researchers from the Louis Pasteur Institute, the 2008 Nobel Prize in Medicine for discovering the virus that causes AIDS. The Nobel citation stated that their HIV discovery was "one prerequisite for the current understanding of the biology of the disease and its antiretroviral treatment. Their work led to the development of methods to diagnose infected patients and to screen blood products, which has limited the spread of the pandemic."[11]

The AIDS research struggle exemplified the complexities caused by ego-driven personalities, avarice, and political arrogance in the struggle for scientific advancement. Unfortunately, the price was paid in human suffering. The net result was that, during a period of a few years, about 10,000 hemophiliacs in the United States became HIV positive. Of those who transfused more than once a month with Factor VIII concentrate, the figure rose to 90 percent.[12] The treatment was the killer.

Fortunately, a tremendous amount of effort went into developing drugs, which, when properly administered, could control, although not cure, the disease. Several years later, a number of drug cocktail combinations were developed that help keep down the viral load and therefore the progression of symptoms. Ironically many of the hemophiliacs avoided the quick death that had been predicted.

There was considerable effort made in both the United States and Canada to provide compensation to AIDS victims for the failure of the FDA and the Canadian Red Cross, which collected and distributed blood and blood products in Canada. But it has not been a timely or facile process. More than 20 years after hemophiliacs accused four American drug companies of intentionally selling them contaminated Factor VIII and IX concentrate, some class-action lawsuits for compensation are still in litigation.

Many patients in the United States and Canada were strongly advised to accept the compensation rather than filing an individual legal action against the manufacturers of the concentrates, because, frankly, the patients would very likely die before the actions could be settled in court. Patients had to consider the impact on their families. There were provisions made for compensation for HIV-positive sexual partners (spouses) of hemophiliacs and to others who, for example, received tainted blood in treatment of other conditions, such as open heart surgery. But it was a long and difficult struggle.

When Adam filed for compensation, he was informed that the tattoo on his arm was the likely source of his HIV infection, not the Factor VIII concentrate that he transfused prior to the viral antibody blood test. Eventually he received compensation for his AIDS infection, but his struggle to prevail against multimillion-dollar corporations was not without emotional and physical consequence.

Adam is one of 33 million people worldwide living with HIV/AIDS. His infection is inextricably linked to the failure of the blood banks and Factor VIII concentrate producers to recognize that the contamination of their blood products was not only possible, but highly likely. The quality of every facet of his life—socially, emotionally, professionally, and financially—has been impacted by his constant struggle to survive with multiple disabling conditions. His parents carry the guilt of having administered contaminated blood products to their son, as well as the rage that comes from knowing that it could have been prevented. They will be forced to live with this anguish for the rest of their lives.

POLITICS

Political considerations have often gotten in the way of new directions for society. Should laws be passed that prevent major employers and government bodies from reducing the profits of companies who donate much to political coffers? Should power companies be forced to reduce their emissions even though it will require big expenditures?

In contrast, many political bodies want to make hay with voters by seeming to be looking out for society's best interests. The Eugenics Movement was politically fueled by the desire to eliminate defective human traits. The methyl-mercury scandals discussed in Chapter 10 are typical of what happens when corporations and governments connive to suppress damage being done to people and the environment. Global warming is far more of a political movement predicting gloom and doom, if action isn't taken now, than a scientific one. All the scientists supposedly agree that man and the emissions of carbon dioxide are the reasons for imminent climate changes, but the truth is far different, and far more careful consideration is needed.

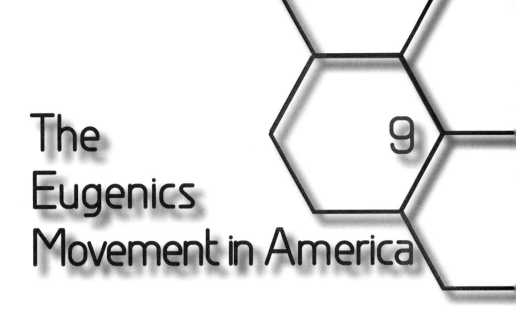

The Eugenics Movement in America

9

At the turn of the 20th century, new ideas were proliferating across the Atlantic from Europe to America's intellectual elite and finding acceptance in the heartland. Austrian priest and scientist Gregor Mendel's laws of plant heredity had just been rediscovered. Biologist and philosopher Herbert Spencer's theory of evolution had been expounded upon in his 1862 book, *Developmental Hypothesis*, preceding Darwin by seven years. He had coined the term "survival of the fittest" and was the father of what came to be known as social Darwinism. English naturalist Charles Darwin's *Origin of Species* was published in 1859, espousing his theory of natural and sexual selection. English explorer, anthropologist, and eugenicist Sir Francis Galton's version of social Darwinism was spurring the imaginations of the American scientific community.

Charles Davenport, a Harvard-educated biologist and instructor of zoology at Harvard University, was the driving force behind American social Darwinism and the Eugenics Movement in America. He was a member of the National Academy of Sciences and highly respected for his 1911 book, *Heredity in Relation to Eugenics*, which became a popular college textbook, and his seminal 1929 "scientific racism" study, *Race Crossing in Jamaica.* In his position as director of the Brooklyn Institute of Arts and Sciences' Biological Laboratory, Davenport sought to improve the human race through better breeding. With funding from the Carnegie Institute of Washington, he established a eugenics research laboratory at Cold Spring Harbor, New York.

Davenport developed his theory of "scientific racism" after studying Gregor Mendel's laws of plant heredity, which were based upon 10 years' meticulous study identifying dominant and recessive inheritable traits in peas. Mendel had carefully charted dominant and recessive inheritable traits in more than 10,000 cross-fertilized pea plants, noting the results of controlled pollination in tall and dwarf pea plants and in smooth pods versus wrinkled pods. Whereas most scientists were debating the efficacy of the cross-application of Mendel's laws in the early 20th century, Charles Davenport, apparently devoid of empathy for those who differed from himself, advocated enforcing Mendel's laws among his fellow Americans. In a 1912 letter to a colleague he declared, "I may say that the principles of heredity are the same in man and hogs and sunflowers."[1] Apparently peas also applied.

He briefly visited with Sir Francis Galton, Charles Darwin's cousin, during a four-month evidence-gathering trip to Europe, and received Galton's warning that the study of eugenics should be a serious scientific enterprise. Galton was a proponent of the idea that "All creatures would agree that it was better to be healthy than sick, vigorous than weak, well-fitted than ill-fitted for their part in life. The aim of eugenics is to represent each class or sect by it best specimens; that done, to leave them to work out their common civilization in their own way."[2]

The primary difference between Galton's and Davenport's ideologies is that of Davenport's perverse concept of moving positive eugenics to the arena of negative eugenics. Whereas Galton advocated the attempt to encourage healthy, productive classes to proliferate a larger proportion of offspring to the next generation, Davenport proselytized the elimination of imperfect, unproductive classes from the next generation. Galton warned that "we can't mate men and women like cocks and hens,"[3] whereas Davenport supported a mandatory process for eugenic cleansing leading to racial purity and imperialistic supremacy. As the founding father of the American Eugenics Movement, he orchestrated a human breeding program gone so awry that social inadequacy was deemed a genetically carried trait and the common denominator for forced sterilization.

Davenport took his ideas to America and attempted to gain support from medical science, but when he failed, he cast his net to the agricultural community and gathered support for his movement in the American Breeders Association. Created in 1903 by the Association of Agricultural Colleges and Experimental Stations and supported by the U.S. Secretary of Agriculture, it had successfully applied Mendel's laws of plant heredity to agriculture. Genetically altered seeds and livestock could produce larger, more productive yields and increase profits in the farming community.

He was familiar with the modern perfectionism ideology of New York State breeder and founder of the Oneida Community, John H. Noyes. Educated at Dartmouth and Yale, Noyes believed that man could achieve a utopian society on earth. He had already advocated selective human breeding in his 1872 paper, "Essay on Scientific Propagation." Charles Davenport knew he had friends in the agricultural community when he read Noyes's belief that "Every race-horse, every straight-backed bull, every premium pig tells us what we can do and what we must do for man... The results of suppressing the poorest and breeding from the best would be the same for them as for cattle and pigs."[4] Although Noyes was deceased, his ideas had found their way to New York's agricultural elite. The association members were eager to embrace Davenport's selective human breeding proposal and to export his scheme to every state throughout America.

By 1909, the American Breeders Association had formed subcommittees for a variety of hereditary human defects, including "insanity, feeblemindedness, hereditary pauperism and racial impurity, called race mongrelization."[5] The first step on the road to the elimination of undesirables was to systematically register the positive and negative inherited traits of every American. Special attention was given to defective human traits.

Davenport seized upon the idea of establishing a eugenics records office for the purpose of quietly documenting the family pedigrees of every American. By doing this, he could systematically, and allegedly scientifically, separate and exterminate defective "germ plasms," what we now think of as genetic traits, from desirable human qualities. But first he would need to solicit funding to carry out his maniacal plan.

He scoured New York's list of wealthy philanthropists for financial support and found it in the widow of railroad magnate E.H. Harriman, Mary W.A. Harriman. Her late husband had already sponsored a Darwin-style expedition to Alaska, organized by two Eugenics Movement sympathizers, and had arranged a meeting between Davenport's circle and President Theodore Roosevelt. Sympathetic to Davenport's cause, she funded the establishment of a Eugenics Record Office in Cold Spring Harbor, New York. Davenport took the helm of three eugenics agencies: the American Breeders Association, the Eugenics Office at the Carnegie Institute, and the newly formed Eugenics Records Office at Cold Spring Harbor.

His next step was to find a co-conspirator with unbridled passion to carry out his devious plan. Harry H. Laughlin fit the bill. He had graduated from the First District Normal School and worked as a high school teacher and principal before he found a position at the Department of Agriculture, Botany, and Nature at his alma mater, the First District Normal School in Kirksville, Missouri. Sympathetic to eugenics principles, Laughlin had already disseminated his treatise for a world government headed by six genetically superior nations. Although he failed to garner support among heads of state, he embraced all of the elitist and racist qualities Davenport desired.

Charles Davenport appointed Harry H. Laughlin to head up the Eugenics Record Office at Cold Spring Harbor. Almost immediately, he hired a dozen field workers to quietly and systematically collect the records on prison inmates, patients in psychiatric hospitals, institutions for the epileptic and feebleminded, schools and homes for the deaf and blind, hospitals, and almshouses. They scoured the East Coast, constructing pedigree charts on families they deemed unfit. The physically disabled were grouped together with the cognitively challenged, the deformed, the criminal, and the economically disadvantaged to create a very loosely defined category of defectives. When one family member met the field workers' criteria for inclusion, the entire family became part of the targeted group.

Laughlin worked tirelessly to promote Davenport's plan to segregate what he termed "the socially inadequate." He promoted a policy of isolating those maintained at the public expense, during their reproductive years, in order to exterminate defective "germ plasm" and eliminate future generations of inadequate people. He obtained a list of public officials in every state in the United States and lobbied them to his cause through fear-mongering. Under the guise of science, Davenport promoted the belief that the bottom of society's barrel was reproducing at a faster rate than the upper crust and would soon overtake and outnumber the worthy class, placing undue stress upon the

Harry H. Laughlin, director of Eugenics Record Office at Cold Spring Harbor, New York. Courtesy of Harry H. Laughlin Papers, Special Collections Department, Pickler Memorial Library, Truman State University.

purse strings of America's finest. He failed to mention that although the lower classes' birth rates were higher, the infant mortality rate placed them on equal ground with America's upper classes. In a convoluted survival-of-the-fittest argument, he predicted the eventual extinction of the very people who supported his cause.

In a comparison between socially unfit human females and feral cats and dogs, Laughlin alleged that homeless strains that "consort freely with equally worthless mongrel sort[s]" multiply rapidly and produce an inordinately high rate of defective stock. He advocated the sterilization of the "lower 10 percent," starting with the insane, the criminalistic, and the "hovel type."[6] The hovel type was composed of families on the lower economic scale with a variety of social problems, including incest, alcoholism, epilepsy, criminal behavior, and feeblemindedness. He projected that the chronically ill and the deformed would probably be the first populations annihilated, followed by the epileptic, the criminalistic, and the mentally ill, in that order.

Slithering beneath the underbelly of the Eugenics Movement was an insidious, surreptitious plan to systematically chart the pedigree of every American family for the purpose of identifying and eliminating future generations of "defective" families. Heavy propagandizing and lobbying by the Eugenics Record Office had already convinced some state legislatures to pass sterilization laws. The State of California led the way by ordering the sterilization of 268 patients incarcerated for mental illness. Of them, 40 were under age 19, and of those one was less than 10 years old.[7]

Davenport proselytized his ideas at the 1914 Race Betterment Conference in Battle Creek, Michigan. It was sponsored by fellow eugenicist Dr. John Harvey Kellogg, superintendent of the Battle Creek Sanitarium and brother of the Kellogg Cereal Company's magnate. A quote by Herbert Spencer, the father of social Darwinism, was emblazoned upon the cover page of the proceedings: "To be a good animal is the first requisite to the success in life, and to be a Nation of good animals is the first condition of national prosperity."[8]

Dr. Kellogg demonstrated his impassioned support of Davenport's fear-mongering ideology with the statement "Man has improved every useful creature and every useful plant with which he has come in contact—with the exception of his own species."[9] He argued that man is influenced by both hereditary and environmental factors, and, according to the laws of evolution, we are as likely to "degenerate as to progress."[10] Next, he included several examples of mankind's failure to evolve into a higher species by comparing the inferior traits of late 19th-century man to the well-developed bodily structure and superior mental capacities of the ancient Greeks. He cited the steady increase in England and Wales of persons who are "unwilling or unable to subsist of their own efforts."[11] He then quoted Dr. Tredgold, the Registrar-General of England, as follows: "Life on this planet is so constituted that it can only progress by the survival and propagation of the biologically fit and the elimination of the unfit."[12]

Kellogg cited the view that the race, like the individual, has only a limited store of vitality and that the growth and decline of the individual resembles the stages of evolution. He relied heavily upon the work of the leading biology professor from Boston University, the late Alpheus Hyatt, who warned that what was earlier attained as the pinnacle of progress will be followed by a stage of decline, ending in extinction. Dr. Kellogg argued that through the study of the laws of eugenics, man must combat race extinction and the natural laws of physical, moral, and mental degeneracy. Within a century, he said, through scientific breeding and the principles of healthful living, we could create a new race of exceptional men. He advocated the establishment of contests that would award prizes to the finest families with the best health and endurance records.

At the same conference, Davenport praised the American Breeders Association's success in producing heartier carnations and chrysanthemums and pointed to its great success in heat-resistant watermelons and disease-resistant plants and animals. He asked, "If such great achievements can appear in the application of the laws of heredity to plants and animals, why cannot something be done in the case of man?"[13] Arguing from the position of the heavy economic burden that the undesirable classes place upon the state, he promoted the idea that crime and poverty, as well as epilepsy, mental illness, and feeblemindedness, are all inheritable traits. He was particularly concerned with the wayward children, especially girls, who were bred from defective stock and would produce a whole new generation of defectives, doubling the burden upon the state. Pockets of degenerate communities, he warned, through defective germ plasm, breed "a stream of people who constitute certainly a large proportion of the paupers, beggars, the thieves, burglars and prostitutes who flock to our cities."[14]

Within a span of two generations, Laughlin projected, the sterilization of approximately 15 million defectives could be accomplished. He suggested if this plan could be quietly and conservatively introduced, it would most likely succeed. However, if mass sterilization were introduced on a more aggressive level, it would most likely result in failure. The plan was to surreptitiously cut off the bloodlines of families the eugenicists identified as unfit for inclusion in American society. The trick was that it would require widespread public acceptance and legislation. In the interest of "safety," he advocated the deferment of his plan for 10 years to permit adequate propagandizing, the enlistment of additional experts to the cause, and the collection of genetic records on every family in America.

The idea of creating better humans through selective breeding failed to gain wide support among sociologists. Herbert A. Miller, PhD, a sociology professor at Michigan's Olivet College, addressed his audience by stating that, although diametrically opposed opinions had been expressed, no one had resorted to name-calling. It was apparent from the tone of his speech that he privately and rightfully held a few choice names for Davenport and his coconspirators. Miller argued the idea that social problems are bred by environmentally determined ethical qualities. By changing the system that generates social strife, caused primarily by poverty, racial prejudice, and industrial conditions, Miller espoused the idea that all but those who are impoverished as a result of biological pathologies could become productive members of society.

Davenport immediately debunked Miller's argument by citing specific examples of young children who had been removed from their degenerate families and raised in good homes and institutions, only to later exhibit the same degenerate traits of their biological relatives. Certainly there is an exception to every rule, and Davenport provided a list of exceptions as ammunition for his hypothesis that nature, not nurture, determined one's social class and behavioral characteristics. He refused to acknowledge the lasting emotional impact that abuse, neglect, and impoverishment in early childhood can have upon later development. Nor was the trauma of being ripped from the arms of ones' mother and placed in an institution attributed to later degenerate traits.

In an orchestrated plan, Harry Laughlin laid out his ideas for the practical application of the eugenics program, "to purify the breeding stock of the race" through education, legal restriction, segregation, and sterilization.[15] The title of his paper, "Calculation of the Working Out of a Proposed Program of Sterilization," leaves no doubt that his primary goal was to terminate reproduction among his perceived "degenerate classes." Presenting charts and graphs citing statistical information which supported his co-conspirators ideology, he advocated early state custody of the lower levels of society, prior to sexual maturity, for the purpose of sterilization, either voluntary or forced. Shorter periods of incarceration, he stated, would result in the processing of increased numbers of defectives, thus cutting off defective bloodlines at a more rapid rate.

Despite impassioned warnings by many of America's academics and journalists, the movement gained wide acceptance among America's elite and acquired increasingly large endowments from the Carnegie Institute, the Rockefeller Institute for Medical Research, and Mrs. Harriman.

A propaganda campaign by the American Breeders' Association promoted Davenport's quiet and conservative plan to chart the family records of the unfit. It was the second step in Davenport's plan to exterminate defective "germ plasm." It started with the introduction of a "Fitter Family" contest at the Kansas Free Fair in Topeka in 1920. An outgrowth of "Better Baby" contests that had gained wide popularity at state fairs, "Fitter Families for Future Firesides" was founded by Mary T. Watts and Dr. Florence Brown Sherbon, a University of Kansas professor known for her support of the Eugenics Movement.

The first "Better Baby" contest was held at the Iowa State Fair in 1911. Part of the Baby Health Examination movement, children through age four were judged against a national standard for weight and height. A Grade A was awarded to the fittest children. Parents of those who received less than an A were encouraged to work to improve their child's grade for the upcoming year. Before the outbreak of World War I, 40 states hosted "Betty Baby" contests.

Each family submitted an "Abridged Record of Family Traits." Children under age 17 were weighed, measured, and given physical examinations, whereas adults underwent blood and urinalysis testing, and a Wassermann test for venereal disease. Mental fitness and physical agility measures were also administered. Family members were assigned a letter grade, and the contest's winning family received a silver trophy. Runners-up who received a Grade A rating were awarded medals.

The contests, which gained widespread popularity, were open to everyone and were divided into small, medium, and large family categories. Even childless couples could enter. Alongside the equestrian, cattle, swine, sheep, and poultry contests, families who entered were judged upon their physical health, behavioral characteristics, and intelligence. Winning families, most of whom were tall, muscular, and had light-complexioned, Aryan traits, were deemed to possess the greatest potential of producing superior offspring. The contests were enormously appealing to many middle and upper class American families and gained widespread acceptance in American society.

Soon Dr. Kellogg's dream of creating a man who will be "in every way, a bigger man...a real aristocrat...of abounding health and vitality, polluted with no disease of hereditary taint"[16] was in full swing at agricultural fairs throughout America. Thousands of unsuspecting families, both "degenerate" and of "the best stock" had recorded their extended family's positive and defective traits for the Eugenics Movement.

Initially, the second wave of the Eugenics Movement intended to influence young couples who planned to marry to selectively breed healthier, more productive children. Those who possessed undesirable traits, such as mental, physical, or socioeconomic inadequacy were urged to reconsider their marital choices, while those who exhibited desirable traits were encouraged to propagate the species by having large families.

When it became apparent that human emotions trumped rational considerations, resulting in offspring who possessed undesirable traits, eugenicists advocated the wider segregation of defectives during their reproductive years. This was carried out through an expansion of the definition of "defective classes" and the involuntary incarceration of perceived "defectives." High priority was given to epileptics, who were lumped into a broad category, including not only those who suffered from seizures (inherited or as the result of brain injury), but also migraine headaches and fainting spells, whatever the cause. Soon epilepsy and feeblemindedness were biologically linked through misinformation as conditions resulting from a common defect of the nervous system, and feeblemindedness received a broad, selectively encompassing definition. New defective categories targeted not only the deaf but the hearing impaired, not only the blind but the vision impaired. Orphans, the homeless, and the impoverished soon found their names on the Eugenic Record Office's list of those possessing defective "germ plasm."

Harry Laughlin sought to eliminate the lower 10 percent of human stock because, in his mind, they were so "meagerly endowed by nature that their perpetuation would constitute a social menace."[17] To accomplish his goal, he would have to successfully lobby states across the entire country to pass eugenical sterilization laws. With the passion devoted to his 1914 speech, he and his cohorts gained political success during the 1920s and 1930s. This was in part because, in 1922, he proposed a "Model Sterilization Law" to prevent the procreation of persons of socially inadequate or defective inheritance.

By 1924, at least 3,000 people throughout the United States had been involuntarily sterilized, mostly in California. Thousands more had consented to voluntary sterilization as a condition for release from institutions to which they had been committed. By the 1930s, more than 30 states had passed eugenical sterilization laws. By the end of 1934, 20,000 forced sterilizations had been performed, mostly in California and Virginia. In all, 27 states enacted eugenics laws.

Elsewhere in America, the burgeoning immigrant population presented a special problem to the eugenicists. They had successfully lobbied the federal government for the exclusion of families whose biological traits were not compatible with good citizenship from entering the country.

Psychologist Henry Goddard had introduced his version of the French psychologist Alfred Binet's intelligence test (designed to measure the increase in the functional level of mentally retarded children by reshaping their environment), to Ellis Island as an indicator of biologically based low intelligence among immigrants. It contained culturally biased questions such as the color of gemstones and the location of America's most prestigious colleges. By implementing this measure, administered in English, Goddard obtained supposed confirmatory evidence that these targeted immigrant groups were racially inferior.

Initially, he selected the dark-complexioned Irish, southeastern Italians, and Eastern European Jews. Later, he concluded that 80 percent of the Jewish, Polish, Italian, and Russian immigrants were mentally defective or feebleminded—the result of defective germ plasm. With funding from Mary Harriman, New York charitable organizations selected Jewish, Italian, and other targeted immigrant groups for deportation, confinement, or forced sterilization.

The American Eugenics Movement took its negative message overseas to the First International Congress on Eugenics in London in 1912. As it gained momentum in Aryan Northern Europe and Germany, American eugenicists contributed intellectual guidance to the movement. Funded by the Rockefeller Foundation, Harry H. Laughlin's Model Sterilization Act became the model for Germany's Hereditary Health Law in 1933, the law that led to the Holocaust. The text in Germany's 1933 law was nearly identical to Laughlin's model. For his contributions and pioneering work, Laughlin was awarded an honorary doctorate from the University of Heidelberg.

The "unfortunate absence" of the German delegation from the 1932 Third International Congress on Eugenics was bemoaned by Charles Davenport. Although the Rockefeller Foundation had contributed vast sums of money to German eugenics research, there were insufficient funds to finance the German delegation's trip to New York. Grants from the Rockefeller Foundation had financed Josef Mengele's early eugenics work,[18] but, by 1936, the Rockefeller Foundation was denying funding to eugenics research due to its political unpopularity.

After Charles Davenport's retirement in 1934, officials from Washington's Carnegie Institute began to move against Laughlin's race-betterment policies and Nazi propaganda. But it was not until December 31, 1939, that the Carnegie Institute formally retired Harry H. Laughlin and shut down the Eugenics Record Office.

From 1936 to 1939, Nazi Germany increasingly became a threat to Europe and the free world. Eugenicists in America were distancing themselves from the Eugenics Movement's Nazi sympathizers. As horror stories from concentration camps made their way across the Atlantic, American sentiment was turning against a policy of negative eugenics.

After World War II, during the Nuremburg Trials, Nazi lawyers cited the constitutionality of a famous Virginia test case in their defense of the compulsory sterilization of 2,000,000 people. Harry Laughlin's Model Sterilization Act served as a mock-up for Virginia's 1924 Eugenical Sterilization Act to legalize compulsory sterilizations of "defective persons." Later that year, the statute went before the court in the famous test case Buck versus Bell, and finally on to the U.S. Supreme Court. Carrie Buck, a young Virginia woman declared feebleminded in a sham trial, had her fate sealed on May 2, 1927, when U.S. Supreme Court Justice Oliver Wendell Holmes delivered the Supreme Court's decision: "It is better for all the world, if instead of waiting to execute degenerate offspring for crime, or to let them starve for their imbecility, society can prevent those who are manifestly unfit from continuing their kind. The principle that sustains compulsory vaccination is broad enough to cover cutting the Fallopian tubes. Three generations of imbeciles are enough."[19]

In a trial that can be only characterized as a miscarriage of justice, Carrie Buck fell victim to the full wrath of the Eugenics Movement, and was first institutionalized at the Virginia State Colony for Epileptics and Feeble Minded in Lynchburg, Virginia, and later sterilized. Carrie had been raped by her foster parents' nephew and gave birth to an illegitimate child. For this she was declared feebleminded and the potential parent of socially inadequate offspring. Later, evidence confirmed that Carrie and her daughter were of normal intelligence.

The Third Reich's extermination program can be found in the American Breeder's Association's 1911 study on the best practical means of exterminating defective "germ plasm"—euthanasia. American eugenicists rejected euthanasia due to their perception of America's sentiment against it. However, many mental institutions practiced passive euthanasia through lethal neglect and intentional infection. Other doctors practiced infanticide.

Some American eugenicists continued to promote euthanasia as late as 1942. Foster Kennedy, an American neurologist, advocated the use of euthanasia in his 1942 article "The Problem of Social Control of the Congenital Defective: Education, Sterilization, Euthanasia," published in the *American Journal of Psychiatry*. He wrote, "I believe when the defective child shall have reached the age of five years...that the case should be considered under law by a competent medical board...to relieve the defective...of the agony of living."[20] He believed that the parents' opposition to the state's decision to kill their children should be a matter of psychiatric concern. Fortunately, it didn't become an institutionalized program in America, although some eugenicists championed the cause. In Germany, more than 300,000 psychiatric patients were euthanized.

The American leaders of the Eugenics Movement were never prosecuted for their crimes against humanity. They lived the remainder of their lives in comfort while thousands of their victims were denied the human pleasure of simply raising a family. We can only hope that our society will never forget the horror of their transgressions.

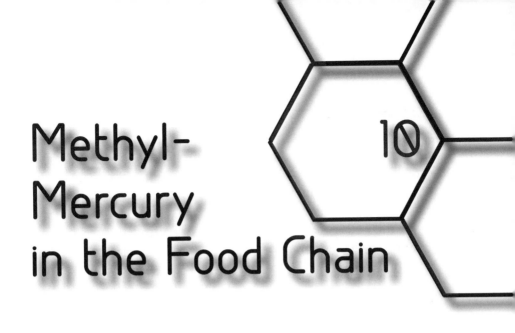

Methyl-Mercury in the Food Chain

10

At the turn of the 21st century, California's San Francisco Bay area was rocked by an outbreak of sub-Minamata disease. The victims were health-conscious individuals who consumed predatory fish such as swordfish, ahi, sea bass, and tuna several times a week.[1] They were not the first to suffer from the disease; a horrific methyl-mercury poisoning epidemic hit Japan in 1956, and killed or maimed hundreds of people.

Minamata disease, or methyl-mercury poisoning, was the natural consequence of the Chisso Corporation's policy of dumping mercury into Japan's Minamata Bay and its tributaries from 1930s to the 1960s. Minamata Bay was once known as a natural spawning ground for fish. The local inhabitants made a living through subsistence and commercial fishing and vegetable crops. Before the arrival of manufacturing, the rural city lived in harmony with its environment, featuring lush vegetation and an abundance of marine life. That changed when Minamata's residents welcomed industry into their village with the hope of economic prosperity.

In 1932, The Nippon Nitrogen Fertilizer Co., Ltd., started developing plastics, drugs, and perfumes through the use of a chemical called

Inorganic mercury is a naturally occurring neurotoxin, which enters the environment through emissions from mining ore deposits, coal burning and waste, and industrial manufacturing plants, as well as through volcanic activity.

Inorganic mercury is converted to its organic form when it is absorbed by small plants and animals in lakes and streams. At each step up the chain, it accumulates in greater concentrations. Predatory fish at the top of the aquatic food chain can have dangerously high levels of the neurotoxin, 10,000 to 100,000 times greater than the water.

acetaldehyde. Inorganic mercury was used as a catalyst in the manufacturing process, and mercury waste products were dumped into Minamata Bay. Nine years later, the company began its first production of vinyl chloride and the local economy prospered as the villagers enjoyed a higher standard of living. It was the sole industry in Minamata and therefore wielded great influence upon the town and its political interests.

In 1950, the company underwent another name change to the Shin Nippon Chisso Fertilizer Co., Ltd. A year later, the company produced more than 50 percent of Japan's total output of acetaldehyde—6,000 tons a year.[2] That same year, the effects of industrial pollution reached a crisis level when fish began to die and float on the surface of Minamata Bay and its contributory waters. It was obvious that something was terribly wrong when dead fish washed upon the shore. The easy supply of fish was a tasty treat for hungry cats and other scavenging animals. Not long thereafter, the village's feline population began to exhibit strange behavior such as convulsions, salivating, disorientation, and suicide by throwing themselves into the bay. The curious and somewhat shaken villagers coined the term "dancing cat disease" after they observed the cats dancing in circles. A humorous sight at first, it became a grave issue when the cats suddenly dropped dead. Birds that fed upon the fish and shellfish all of a sudden started to drop out of the sky. Next, dogs, pigs, and other animals began to exhibit the same fatal neurological symptoms.

In March 1956, a 5-year-old girl who had developed similar symptoms was taken to the hospital owned by the Chisso Corporation. Formerly a very bright and agile girl, throughout a two-month period she became incapable of holding her chopsticks to feed herself. Shortly thereafter, she became unsteady on her feet, salivated excessively, and had difficulty verbalizing her needs. Additionally, she was no longer able to fall asleep. Her physician noted that she appeared to be malnourished and exhibited signs of dementia. She was admitted to the hospital on April 23 with spastic paralysis, insomnia, and convulsions. By May 28, the progressive symptoms riddled her tiny body. As the syndrome progressed, she became blind and exhibited increased convulsions and deformity in her extremities. Her tiny body took on the extreme spasticity associated with severe cerebral palsy.[3] Her vocalizations were sudden, uncontrollable cries and her tongue was punctured with bite marks. One can imagine the widespread panic that ensued in her village when she eventually became comatose and died.

Shortly thereafter, more victims who lived in fishing neighborhoods along Minamata Bay began to stumble while walking and experience numbness in their limbs. More children appeared at Chisso's hospital and were admitted with the same mysterious neurological symptoms. The first victim's parents, along with every other person in their neighborhood, developed symptoms. Their speech became slurred and they could no longer accomplish fine motor tasks, such as eating with chop sticks or buttoning buttons. They developed difficulty hearing, drooled, lost their visual range, and began to tremble uncontrollably. Insomnia was also noted. Many died.

Soon pregnant women delivered babies with terrible disabilities, including mental retardation and physical spasticity. Many died in infancy. On May 1, 1956, Dr. Hajime Hosokawa, Director at the Chisso Hospital, reported that "an unclarified disease of the central nervous system had broken out."[4] By October, 40 patients had been treated at the Chisso Corporation's hospital, of which 14 had died—a mortality rate of more than 35 percent.[5]

No one knew what caused the strange epidemic or how to stop it. Some speculated that it was a contagious viral inflammation of the brain and nervous system, which contributed to the increasing sense of panic. Local physicians suspected that it was a contagious disease, so patients were quarantined in their homes and shunned by the community. Efforts were made to disinfect the living quarters of the victims, but the disease continued to spread. No one could identify the cause of the strange new plague that was spreading throughout the village's fish-eating population.

The city government invited researchers from Kumamoto University Medical School to assist in the investigation. As the research group visited Minamata on a regular basis and examined patients at the university hospital, a more complete picture began to emerge. Household data revealed that most of the victims lived along Minamata Bay and consumed fish and shellfish they harvested from the bay.

It took just two months for the research team to speculate that the disease was generated by the consumption of fish and shellfish and that the Chisso Corporation's contaminants were the most likely cause. However, Chisso's officials immediately denied the charge and continued to do business as usual. When Chisso came under pressure from the public after investigators revealed the company was dumping toxic chemicals into the bay, it brazenly remediated the problem by diverting their waste into the Minamata River Delta, on the other side of town. Within months, Minamata disease followed.

The Chisso Corporation refused to cooperate with the medical school's research team and withheld information on its industrial processes. Instead, it poured money into private research in an effort to provide evidence that its effluent waste was not the cause of Minamata disease.

Employees from the Chisso Corporation served on committees formed to investigate the problem, and in a blatant conflict of interest, debunked as nonsense the idea that there was a direct link between Chisso and Minamata disease. This changed in February 1959, when a team of scientists found large concentrations of methyl-mercury in fish and shellfish in Minamata Bay, and exceedingly high amounts of the toxin around Chisso's effluent canal.

Dr. Hosokawa and his staff at Chisso's hospital conducted an experiment in which they sprinkled effluent waste water from Chisso over some cats' food. At 46 days after the onset of the experiment the first cat began to exhibit signs of partial paralysis in its hind legs. It progressively degenerated into complete paralysis with tremor and was euthanized. When Hosokawa reported his results to the company, he was denied access to contaminated waste water and taken off the experiment. Dr. Hosokawa endured severe criticism by his employer, and his experimental research results were buried in Chisso Company files.[6] This clear evidence of criminal intent by Chisso would not emerge for another decade.

On November 12, 1959, the Ministry of Health and Welfare issued the following statement: "Minamata disease is a poisoning disease that affects mainly the central nervous system and is caused by the consumption of large quantities of fish and shellfish living in Minamata Bay and its surroundings, the major causative agent being some sort of organic mercury compound."[7]

Organized demonstrations by local fisherman at the Chisso factory brought the issue to the public's attention, and the it demanded a purification system to detoxify the effluent flowage. Chisso installed a Cyclator to detoxify mercury emissions, but it was ineffective. It has been reported that the plant perpetrated this charade to appease the public when it was fully aware it was pulling off a cover-up. Chisso also offered a consolation payment of $540 to 120 victims, 20 of whom had died, and threatened job loss unless the families of victims accepted the low amount of compensation. As a condition for compensation, the victims were required to sign a contract that absolved the Chisso

The Shiranui Sea is an inland sea surrounded by the Kushu and Amakusa Islands in southwestern Japan. Minamata is located on the southern coast of the Shiranui. Its ragged coast has many inlets and coves, which provide a perfect environment for the spawning of fish and shellfish. When contaminated with methyl-mercury, they provide a highly toxic diet for larger predatory fish.

Corporation from responsibility for their plight, even if the company was later found guilty. It was hardly just reparation for the loss of life, health, and well-being.

Government authorities and Chisso failed to take remedial action or to warn other companies in Japan that used the same chemical process in its manufacturing of the danger. As a direct consequence of their inaction, there was a second outbreak of Minamata disease in Niigata at the Showa Denko Company. In July 1965, 26 cases had been confirmed and five patients had died.

Despite the new cluster of patients with methyl-mercury poisoning, the Japanese government did not officially recognize mercury pollution as the cause of Minamata disease until September 1968. The announcement was made four months after the Chisso plant stopped using mercury as a catalyst in the production of acetaldehyde because it had become obsolete. It was not until five months later that the government designated the Minamata Sea a protected marine area. As a consequence of official denial, methyl-mercury poisoning continued to claim victims, including infants born with acute neurological disabilities. In an effort to protect Chisso from financial disaster the government urged its Certification Committee to restrict diagnosis to those who exhibited the most severe symptoms of methyl-mercury poisoning. Again, the government placed company profits ahead of human health and well-being when it delayed compensation for congenital victims.

In 1973, Chisso lost its court battle, and the contracts previously signed for consolation payments that restricted compensation were declared invalid. This was due, in part, to former medical director Dr. Hosokawa's court testimony confirming that Chisso knew its waste dumping had caused methyl-mercury poisoning. To ward off financial disaster, Chisso sought, and was awarded, relief from the Kumamoto and federal governments.

The financial implications were immense, so in 1977 the Environmental Agency reverted back to the Certification Committee's restrictive guidelines for a formal diagnosis and compensation. More lawsuits ensued and a political solution was finally reached by the federal government in 1995 under the condition that victims would withdraw their lawsuits and agree to a mediated settlement with Chisso.

Victims in central Japan appealed, and in October 2004, won their lawsuit in the Supreme Court.

The Minamata Bay Pollution Protection Office opened in 1976, and, 17 months later, a Sludge Disposal Project began. By 1997, the methyl-mercury level in local fish and shellfish had dropped below minimum safety standards for three consecutive years. Fishing in the bay was reopened for the first time in 24 years.

As a consequence of the extreme social and environmental impact industrial pollution had upon Minamata, the village instituted new guidelines to establish the development of a model city. In 1992, it participated in the Earth Summit World Urban Forum, and its "Declaration on Environment and Development" pledged to adopt the principle of "An Industrial-Cultural City that Values Environment, Health and Welfare." In 1999, as a result of its proactive stance, Minamata attained International Environmental Standards. In 2007, Environmental City, a plan for the revitalization of a healthy Minamata, was accepted by the National Regional Revitalization Plan. Today it has made great strides forward, though it continues to recover from the devastating effects of arrogance, ignorance, denial, and political corruption.

Japan is not the only country whose residents have suffered the harmful effects of methyl-mercury poisoning. Cargill Incorporated, of Minneapolis, and its Mexican subsidiaries had shipped 73,202 metric tons of mercury fungicide–treated wheat and 22,262 tons of treated barley to Iraq in the autumn of 1971.[8] Reportedly, the Iraqi government requested that all grain be treated with the fungicide to prevent mold growth due to humid conditions during transport on container ships. Mercury compounds to prevent the growth of various fungal diseases had only recently been banned in Scandinavia and several American states due to environmental risks, so the price had plummeted, making it a desirable commodity in less environmentally restrictive countries.

Due to unforeseen delays, the sowing season had already passed when the seed grain shipments arrived on cereal-growing farms in south, central, and northern regions of Iraq. Disappointing crop yields during the previous season had left the farming communities in near famine conditions, so to ward off starvation, farming families ingested it in the form of baked bread and fed the grain to their livestock. Soon, birds and rodents near the grain storehouses perished after consuming the seed, and within months there was an outbreak of what was at first suspected to be a viral inflammation of the brain and nervous system among the farmers and their families. Intensive investigation following the Minamata disaster in Japan and the environmental hazards in the United States and Europe had alerted the scientific community to the methyl-mercury poisoning problem.

Iraq's central government acted swiftly and ordered farmers to relinquish all treated grain within a two-week period, under penalty of death. News traveled slowly across farming communities, until the Iraqi army began to carry out its execution order for those who remained in possession of the fungicide-treated grain. Consequently, the frantic farmers disposed of their grain stores in rivers, canals, and throughout the countryside, causing further contamination of birds, fish, and groundwater.

Whereas blood methyl-mercury levels averaged less than 10 per milliliter in the rural population, blood levels were as high as 4,000 per milliliter in those who had consumed contaminated bread.[9] When methyl-mercury passed through the placental barrier to the fetus in pregnant women, infants had even higher blood levels than their mothers. These higher levels persisted for months, and the devastating neurological effects, including mental retardation, blindness, and spasticity, will unfortunately afflict the tiniest victims for the remainder of their lives.

Clinical studies documented tremor, hyper-salivation, unsteadiness, muscle weakness, tunnel vision and blindness, tinnitus and deafness, unintelligible to staccato speech, impaired memory, comas, and anxiety and depression in the affected adults and children.[10] In all, 6,530 victims were hospitalized with acute symptoms, and 459 died.[11] Of approximately 100,000 people poisoned by methyl-mercury in Iraq, it is estimated that 10,000 died.

Japan and Iraq are extreme examples of the consequences methyl-mercury poisoning has on unsuspecting populations. Not only did it kill and maim thousands of humans, but it also destroyed marine life and land animals at each step up the food chain.

Although mercury is ubiquitous throughout our natural environment, human activity has pushed it above safe levels. The burning of fossil fuels, cement kilns, petroleum refineries, chlorine-producing factories, treated sewage, medical waste, florescent light bulbs, mercury thermometers, landfill seepage, and industrial waste products all contribute to the problem.

The sub-Minamata patients in the San Francisco Bay area, mentioned at the beginning of this chapter, were initially presumed to have been contaminated by mercury used to extract gold ore during the California Gold Rush and by mercury mines in the Coast Ranges. An estimated 26 million pounds of mercury was used to extract ore in Northern California, of which 3 to 8 million pounds polluted the environment.[12] California's Sacramento River, which contains several species of fish with dangerously high methyl-mercury levels, discharges into the San Francisco Bay. The level of pollutants varies according to weather conditions, with the highest levels of toxic discharge following storm water run-off.

California's effort to clean up its environment has faced decades of delays caused by litigation. In 2004, California state officials issued a plan to reduce toxic metal emissions in storm water run-off and from the defunct gold and mercury mines by requiring cities to cut mercury releases by 40 percent during a 20-year period. On February 12, 2008, the Environmental Protection Agency approved an amendment which establishes guidelines for a total maximum daily load of mercury into San Francisco Bay. Scientists estimate that the bay's mercury level will return to pre–gold rush quality in approximately 120 years. But there are additional factors that are not easy to regulate; for instance, mercury pollution from Asia travels through the atmosphere and falls upon California's land and watersheds. This makes it a worldwide problem—a problem with no easy solution.

The Great Lakes, the world's largest freshwater ecosystem, have historically been a dumping ground for toxic mercury. In 1970, pike and pickerel from Lake St. Clair tested at as much as 24 times the maximum level deemed safe for human consumption.[13] As a result, local officials banned fishing on the lake and on part of Lake Erie, and many large-scale fisheries were forced to close. Ontario officials gave 11 paper and chemical companies a deadline to clean up their pollutants. Canadian officials discovered Dryden Chemicals Limited in Northwestern Ontario had dumped more than 20,000 pounds of mercury-contaminated wastewater into the English-Wabigoon River system between 1962 and 1970, which eventually contaminated Lake Winnipeg.

The Indigenous People on the Grassy Narrows and White Dog Reserves, who lived just downstream from the plant, and subsisted on fish from the river, suffered greatly through economic devastation and the consequences of Minamata disease. Many were relocated to safer living quarters but were not informed about the cause of their illness until the 1980s.

Chronic mercury exposure can have a serious impact upon fertility and the outcome of pregnancy. It interferes with the part of the brain that controls reproduction and results in menstrual cycle disorders. In men, organic mercury can cause low sperm count, minor genetic damage, a reduction in libido, and impotence. It has also been linked to an increased level of cardiac arrhythmia and heart disease, autoimmune disorders, kidney disease, and liver disease in both men and women.

A 1999 Health Canada report indicated that between 1971 and 1996, 17,671 Indigenous People were found to have dangerously high levels of methyl-mercury in their blood.[14] A 2004 study indicated that there are still high concentrations of methyl-mercury in pike, walleye, and otters in the English-Wabigoon River system. Additionally, residents of all three communities continue to suffer from the devastating effects of the deadly neurotoxin.

The other main source of mercury pollution in Canada was the Dow Chemical Company plant on the St. Clair River in Sarnia, Ontario. Its chlor-alkali operation released more than 22,000 pounds of mercury per year into the effluent flow until 1970, when a pollution-control system was mandated. However, it was not until 2004 that Dow Chemical completed the final phase of its sediment remediation project at Sarnia.

A study done by Health Canada in the 1990s and leaked to the public in 2000 identified 17 areas of concern. The study found a series of outbreaks of Minamata disease in Thunder Bay, Collingwood, Sarnia, and Cornwall. Each of the affected areas had large mercury cell chlor-alkali plants. The Aamjiwnaag Reserve near Sarnia, which is surrounded by 46 chemical plants and refineries, experienced a 40-percent drop in male baby births in the mid-1990s. Of particular concern is the devastating effect of methyl-mercury, and nearly 500 other chemicals, on babies born to mothers who eat Great Lakes fish. One study showed that children exposed to these chemical pollutants have a higher risk for developmental problems and learning disabilities. Another study concluded that these children are as much as two years behind their peers in reading and math.[15]

Prevention on behalf of the International Joint Commission was completed in August 2007. The U.S. Centers of Disease Control and Prevention, in cooperation with the joint commision responsible for overseeing issues related to the Great Lakes, completed a study in August 2007. However, it was not immediately released. It was made public in February 2008 after it was leaked to the Center for Public Integrity in Washington, D.C. It identified 26 areas of concern and revealed that at least nine million people in the Great Lakes basin are at risk for elevated levels of illness and infant mortality due to mercury pollution and methyl-mercury bioaccumulation.

Coal-fired power plants were implicated as a major contributor of environmental mercury pollution when Sweden, Canada, and the United States, despite measures to reduce mercury pollutants during a 20-year period, discovered elevated levels of methyl-mercury in remote lakes with no known nearby source of mercury contamination. Mercury vapors in the atmosphere are dispersed far and wide and can circulate for years. Acid rain, produced by vapor from coal-fired plants, has increased the acidity of rivers and lakes and created a more hospitable environment for the production of methyl-mercury than in more alkaline lakes. In turn, aquatic life has developed increasingly higher levels of methyl-mercury in its tissues. A 1988 study showed that Sweden had more than 9,400 lakes containing fish with unacceptable levels of methyl-mercury. During the same time frame, nearly 70 percent of the samples collected from 67 Michigan lakes exceeded acceptable levels. Similar levels of methyl-mercury were discovered in fish harvested from lakes in Minnesota, Wisconsin, and Ontario, Canada.

In the winter of 1997–98, the Environmental Protection Agency released to Congress two reports on mercury pollution. The 1997 Mercury Study Report identified fossil fuel–fired power plants as the largest source of human-generated mercury emissions in the country. The 1998 Utility Air Toxics Report identified mercury as the toxic pollutant of greatest concern in the United States. The EPA was charged with the task of proposing regulations to control mercury emissions by the end of 2003. It found that there are cost-effective ways of controlling mercury emissions, the technology is available to do so, and improved emerging technologies were being developed. The good news is that mercury emissions could be reduced to safe levels at a cost less than 1 percent of the industries' revenues.

The bad news is that virtually every state in the United States is at risk for methyl-mercury toxicity. In 2005, 500 coal-fired power plants in the United States supplied half of our total energy. They spew out approximately 48.3 tons of mercury into the air each year, an increase of 1 percent since the year 2000. The electric power industry has plans to build 153 additional coal-fired plants. Airborne mercury particulates fall upon our land, our homes, and our lakes, rivers, and streams. It takes only 1/70th of a teaspoon of mercury to contaminate a 25-acre lake. When the methylization process begins, the poisonous organic mercury makes its way up the food chain, increasing in toxicity as it moves closer and closer to human consumption. More than 40 states have issued warnings against consuming fish from lakes and streams. More than half of those warnings apply to all bodies of water in the state.[16] Despite the warnings, many residents are not aware of the danger.

The primary dilemma lies in the fact the coal is readily available as an inexpensive energy source. More than half—52 percent—of our electricity is generated by coal-fired burners, and each household uses approximately 9.5 tons of coal per year. To date, the United States has no comprehensive monitoring system for mercury emissions. Additionally, methyl-mercury contamination across the United States is uneven; a number of characteristics, including precipitation, prevailing winds, landscape, elevation, and proximity to coastal areas and to coal-fired plants, play a role in mercury contamination.

Methyl-mercury poses the greatest risk to children. Because the placenta transports methyl-mercury to a developing fetus, it can cause permanent damage to the developing brain and nervous system. Methyl-mercury halts cell division in the fetal brain and has a profound impact on the cerebellum, the control center for balance and muscle coordination. As many as 630,000 children born annually in the United States are at risk for life-long neurological problems, including learning disabilities, motor deficits, attention and language deficits, and impaired memory, vision, and motor functions.

New Hampshire's Merrimack Station Clean Air Project. New Hampshire's biggest toxic emissions polluter is constructing a new smokestack and scrubber. Photo by Kathleen Marden.

It is estimated that 15 percent of women of childbearing age in the United States have methyl-mercury blood levels that pose a risk to a developing fetus. In March 2004, the U.S. Food and Drug Administration and the Environmental Protection Agency issued a joint warning that restricted the total safe intake of toxic seafood for women and children to no more than 12 ounces per week, including no more than six ounces of albacore tuna. The warning advised zero consumption of shark, swordfish, king mackerel, and tilefish due to their high methyl-mercury levels.

Clinical studies have found a positive association between a higher fish intake and better cognitive performance in young children, however, there was an inverse relationship between higher mercury levels and cognitive performance. The key is in consuming fish with vital nutrients and low levels of methyl-mercury, but government agencies have bent under pressure from commercial fishing conglomerates.

An EPA letter dated December 5, 2008, warned that the FDA ignored scientific studies showing that species vary widely in their accumulation of methyl-mercury. Scientific studies show that women who ate more than the two recommended servings of fish per week had concentrations of methyl-mercury in their blood seven times the levels of non-fish eaters. Efforts need to be made to educate the public about the dangers of eating fish high in methyl-mercury and the benefits of eating safe fish, such as lobster, tilapia, oysters, salmon, haddock, and herring.[17] This table lists seafood with lower mercury levels:

Lower Mercury Levels in Commercial Fish and Shellfish (Updated 2006)

Species	Mercury Concentration PPM		
	Mean	Median	No. of Samples
Lobster	0.09	0.14	9
Tilapia	0.010	ND	9
Oyster	0.013	ND	38
Salmon (fresh/ frozen)	0.014	ND	34
Hake	0.014	ND	9
Haddock	0.031	0.041	4
Crawfish	0.033	0.035	44
Pollock	0.041	ND	62
Anchovies	0.043	N/A	40
Herring	0.044	N/A	38
Mullet	0.046	N/A	191

Data Source: FDA 1990–2004. FDA made technical changes on 2/8/2006.

Courtesy of the EPA.

Although the Clinton EPA proposed reducing mercury emissions from fossil fuel–fired power plants by 90 percent by the end of 2007, the environmentally indifferent Bush EPA reversed its course. In 2004, as part of its Clear Skies legislation, it proposed a 70 percent reduction in mercury emissions by 2018. This decision will allow an additional 500 tons of mercury to enter our environment and wreak havoc upon not only men, women, and children, but also our aquatic populations and those species that depend upon them for sustenance.[18]

In an effort to satisfy the corporate interests of the coal-burning utilities, the Bush EPA deceptively determined that the emissions data from the top-performing 12 percent (according to law) of the coal-fired plants was not adequate to determine the Maximum Achievable Control Technologies. Therefore, in an act of public deception, it instructed its scientists to increase the number of emission points to calculate the MACT, working backward from a politically calculated figure. Further, it denied that any mercury-specific control technology had been installed in any of the plants, falsely claiming that they were not commercially available. The truth is that particulate scrubbers, activated carbon injectors, and methods of catalytic conversion had been developed and were available, although they may not have been as effective as mercury-scrubbing technologies.

According to the EPA Inspector General's report, senior management instructed EPA staff to develop a Maximum Achievable Control Technologies standard that would place national emissions at a predetermined level of 34 tons per year. By achieving this figure, the officials were able to deceptively place coal-fired utilities in the "cap and trade" category—a category normally reserved for non-toxic pollutants and multi-pollutants. Although mercury is in the same toxic pollutant category as lead, this political manipulation effectively placed it in a less injurious category. It allowed some plants to trade mercury pollution credits with other less polluting plants, creating an unequal toxic distribution. Thus some areas of the country would become hotspots for mercury and methyl-mercury contamination, whereas others would bear less of the burden from environmental contamination.

In March 2005, nine states (New Jersey, California, Connecticut, Maine, Massachusetts, New Hampshire, New Mexico, New York, and

Vermont) sued the EPA as a result of this assault upon the public's right to be protected from harmful mercury emissions. Early in 2008, the United States Court of Appeals for the District of Columbia ruled in the public's interest when it determined that the EPA had unlawfully decided to remove power plants from the most protective requirements of the Clean Air Act. The Bush administration's EPA caused a significant delay in major efforts to reduce dangerous levels of this toxin in our environment.

Methyl-mercury poses a worldwide threat to the health and welfare of our environment and the flora and fauna that flourishes in it. We can no longer afford to be complacent and permit corporate interests to wreak havoc upon us. Science and history have demonstrated the devastating consequences of political manipulation and corporate cover-up. On-site state-of-the-art pollution equipment to control toxic substances, including mercury, is technically possible, and, according to the Government Accountability Office, affordable. So far, 18 states have enacted laws or regulations to control emissions at coal-fired power plants. With honesty and diligence, nothing is impossible.

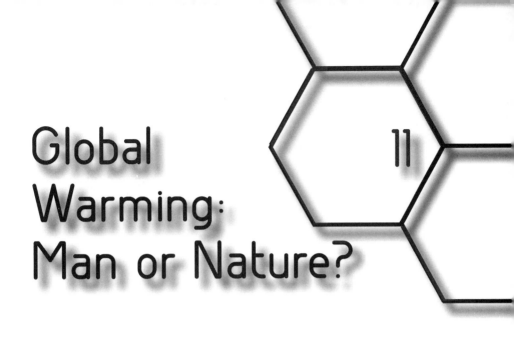

Global Warming: Man or Nature?

11

Rarely has a subject received so much attention as has the notion of "global warming," especially since the publication of Al Gore's *Inconvenient Truth*, the Nobel Peace Prize received by him and the IPCC (UN Intergovernmental Panel on Climate Change) in 2007, and the media hype. If one were to believe the propaganda, CO_2 (carbon dioxide) is public enemy number one. Its increasing production by the world is leading to disastrous consequences, and hundreds of billions of dollars must be spent as soon as possible to reduce the warming and all the damage that will be accompanying it. Use of fossil fuels must be reduced or eliminated. Countries must sign agreements to reduce their emission of carbon dioxide no matter what it costs. Higher-performance cars must be devised. Full subsidies must be given for solar and wind power. If these measures aren't taken, then, in the words of Chicken Little, "The sky is falling."

Although there is nothing simple about predicting the weather or evaluating the myriad of statistics available about it, here are some of the assumptions on which the calls to action are based:

1. All scientists have reached a consensus that Gore and the IPCC are correct.

2. The world is rapidly heating up.

3. The major cause of the supposedly increasing temperature is mankind's increasing production of evil CO_2. Anthropogenic Global Warming (AGW), which is caused by people, is to blame, and Mother Nature is innocent.

4. Action must be taken immediately or we are doomed.

5. Primary threats include rising of the world's ocean levels by as much as 20 feet as a result of the melting of various glaciers, especially on Greenland and in the Antarctic, leading to a huge loss of lives and habitats for residents of low-lying coastal areas, such as Bangladesh and Manhattan.

6. An increased number of very destructive hurricanes, cyclones, tornados—all as a result of global warming—will occur.

7. Polar bears are decreasing in number because of the melting ice, and they need to swim greater distances to find food.

8. Islands, such as the Maldives Southwest of India, are slowly sinking as the ocean rises.

As it happens, in the real world, all of these assumptions are seriously being called into question by a growing number of so-called deniers. Though still difficult, it has become easier to publish papers that seek to replace widely held myths with facts in refereed scientific journals. A turning point may have occurred when BBC News published an article by Paul Hudson in October 2009 entitled "What Happened to Global Warming?" The BBC had previously been fully behind the "Kill CO_2" movement, but now Hudson noted that for the last 11 years we have not observed any increase in global temperatures (and that the global climate models did not forecast it), even though man-made carbon dioxide, the gas thought to be responsible for warming our planet, has continued to rise. Note that, simply put, the temperature of the world has *not* risen for 11 years.

Hudson noted that according to research conducted in November 2008, by Professor Don Easterbrook from Western Washington University, the oceans and global temperatures are correlated. He says that they warm and cool cyclically. The most important cycle is the Pacific Decadal Oscillation (PDO). This was in a positive cycle (warmer than usual) for much of the 1980s and 1990s, and global temperatures were warmer too. In the past, the cycles have lasted for about 30 years, with the period from 1945 to 1977 coinciding with one of the cool Pacific cycles. Now it is again in a cooling mode. In September 2009, Dr. Mojib Latif, a prize-winning German meteorologist and oceanographer, and a member of the Intergovernmental Panel on Climate Change, wrote that we may indeed be in a period of cooling that could last another 10 to 20 years. The current level of CO_2 is about 380ppm (parts per million). Some believe that we must do all we can to bring it back to 350ppm.

Everybody knows that the weather changes from day to day and season to season, and that even local forecasts a day in advance can be wrong. The notion that the "sophisticated" computer models used to predict climate change over the next 20 to 50 years for our entire planet are accurate is mind boggling. Weather depends on many factors, including what happens in uninhabited land areas and over the oceans, which cover about three quarters of the planet. There are external factors, such as sunspots, cosmic rays, variations in the energy output of the sun, El Niño, Ocean Decadal motions, volcanic eruptions, and pollution of the atmosphere, about which we are slowly beginning to learn. It seems strange, but we sometimes forget that CO_2 is not the most abundant greenhouse gas. Water vapor has a much higher concentration of CO_2. (Evaporation of water is also dependent on many factors, which include cloud cover, changes in the surface characteristics of water and ice, winds, and temperature.) CO_2 also comes from the chlorofluorocarbons that were used in spray cans, methane from farm animals, exhausts from jet aircraft, radio waves bouncing off the ionosphere, ozone, and so on. In addition, we haven't had many, many years' worth of good data as to what the actual surface temperatures have been across the planet. Satellite measurements can be very useful, but obviously haven't been available for many decades.

There is considerable concern about the effects of local factors on the surface measuring devices, especially near cities, which tend to hold in heat, and even near structures out in the country. Some devices have even been moved to different locations, though that fact was not noted in their regularly reported periodic temperature compilations. Many devices previously located in the country have been moved to airports.

There are some indirect means of trying to get a handle on past temperatures, such as the use of tree rings (dendrochronology). Cores are taken and the thicknesses of the tree rings each year have a close, but certainly not perfect relation to the overall temperatures that year at the location of the tree. However, not surprisingly, tree rings and tree growth are influenced by other factors besides temperature. Rainfall, shade, and root nutrition are among these. One almost bizarre example of the difficulty of using dendrochronology was when a dozen trees at one location were used (even though many more had been examined). Though all were growing in the same area, the results were nowhere near identical. There was also some indication that the data was cherry picked, so that only those trees giving certain results were used. This is not surprising for propaganda and politics, but is surely not the way of science.

There should be no surprise that politics has been such an important part of the global-warming warnings. Al Gore, after all, is not a scientist, but a politician. Likewise, the IPCC is a much-politicized body—the members are supported by their home governments. The actions that are being discussed all involve politics: How much should we reduce our CO_2 production (as though passing a law would accomplish the reduction)? How many hundreds of billions of dollars should be spent on ameliorating CO_2 production? How much should the developed countries give to the undeveloped ones to assist in their attack on CO_2?

Most CO_2 is produced by the burning of fossil fuels in power plants, but no government will tell its people to keep their houses much colder in the winter and turn off air conditioning in the summer, or tell industry to reduce its output, or cities to be darkened. Closing all coal-fired power plants would be disastrous in many places, but some extremists have demanded such an action. One gets votes, after all, by making promises that one hopes will be forgotten once it's seen that they cannot be kept.

Politics has also been very important in determining the awarding of research contracts. The worse the situation is made to seem, the more research must be done. Thus, most publications discussed in the media provide a range of values for how much the temperature or sea level will increase. The focus is always on the high and usually unrealistic end. For example, some have claimed the sea will rise 20 feet; current rates are about a millimeter a year.

Not surprisingly, one doesn't hear much about the benefits of higher CO_2 levels, such as increased plant growth and crop yields which has been demonstrated in controlled experiments. Also, many countries, such as Russia, would prefer to have a warmer climate. There has been great politicizing in what papers get submitted for publication, because contrarians risk losing their jobs or being denied future research grants if they speak out. A polar bear expert, Mitchell Taylor, who had attended a special conference of polar bear experts every year since 1981, had his paper rejected in 2009 because it didn't follow the party line as to how much danger the bears were in from global warming. He wasn't even permitted to attend the conference and was replaced by people who knew nothing about polar bears.

Politically, it isn't acceptable to talk about the fact that water vapor is the most prominent greenhouse gas, much more so than CO_2. However, it is very much more difficult to predict accurately the effect of water vapor on planetary temperatures. Far more of the planet is covered with water and ice than with power plants. When water evaporates into the atmosphere, clouds form, and they are blown by unpredictable winds. Clouds keep some solar radiation from reaching the planet by reflecting it back out to space, thus cooling the planet. However, the clouds also absorb some of the heat emitted by the ground and help heat the atmosphere.

It is certainly clear that there have been warmer periods of time than the present, which could not have been caused by CO_2, because so little industrialization existed then. There have also been lengthy cooler periods, which also obviously had nothing to do with CO_2. The famous "hockey stick" graph shows what seems to be level temperatures for a long time and then a steady increase because of CO_2. More careful and honest work shows that the curve just happens to omit periods of higher and lower temperature that could not have been influenced by the production of CO_2, and has discreetly been left out of recent IPCC publications.

The history of environmental movements certainly includes examples of bandwagon-jumping to take care of a perceived problem, often with severe and unplanned consequences. One of the better examples is the banning of the pesticide DDT in 1972. This was directly the result of the hue and cry stemming from Rachel Carson's 1962 book, *Silent Spring*. According to her, egg shells of predatory birds, such as hawks, had become thinner because of DDT. The problem is that DDT was by far the most effective, inexpensive, and safe weapon against the anopheles mosquito that spreads malaria. Now, because of the banning, there have been literally millions of deaths, especially amongst young children in Africa. One might wonder if this is a fair trade-off.

A much more recent example involves the production of biofuel to reduce the use of imported oil. Producing corn to be converted to biofuel greatly increased the income of farmers, but, unfortunately (though predictably), substantially raised the cost and reduced the supply of food for consumers. In addition, more detailed calculations have indicated that sometimes more production of CO_2 was produced by all the activities associated with the farming and the extraction of the biofuel than would have been produced using the equivalent amount of oil.

In addition, the pressure for non–CO_2 producing (renewable) power plants, such as solar and wind power, has been dependent on major government financing, incentives, and subsidies. Because of these it has been profitable to build large solar and wind facilities, but operating them requires much higher expenditures than using non-renewable resources. It was found in California, which does have abundant sunlight, that people bought solar swimming-pool heaters when substantial tax and subsidy benefits were provided. They stopped when the benefits were eliminated, causing many companies to go out of business, which meant that much repair and servicing of the solar heating systems could no longer be provided.

There are other strange aspects of the anti-CO_2 war. A number of anti-nuclear groups have loudly proclaimed the need to avoid building new nuclear plants and hopefully to shut down old ones. They are also against CO_2. But the nuclear power plants produce far, far less

CO_2 than do any other major sources of power production. Some countries in Europe, such as Germany and Belgium, have recently delayed earlier mandates to close their existing nuclear power plants by 10 or more years because there aren't reasonably affordable alternatives. Somebody—though never the activists—has to pay the bill.

It should not be surprising, considering the examples given in other chapters, that there have been unexpected but significant new scientific developments concerning the factors that control global warming. It was announced on October 18, 2009, that the *New Phytologist Journal* (184:545–551, November, 2009) had published an article, "A Relationship between Galactic Cosmic Radiation and Tree Rings" by Sigrid Dengel, Dominik Aeby, and John Grace, concerning an evaluation of tree-ring growth rates as a function of various parameters, such as temperature and precipitation. It turns out that there was no significant correlation with temperature or precipitation. However, there was a significant correlation with galactic cosmic radiation. All the trees that were used, Sitka spruce, had been planted in 1953 and cut in 2006. Felling protocols had been laid out by Forest Research; North and West directions were marked on the bark and the discs were frozen as soon as returned to the research station. The rings were counted in their frozen state; otherwise discs can shrink and crack. To quote the authors so as not to bias the reporting: "There was a consistent and statistically significant relationship between growth of the trees and the flux density of galactic cosmic radiation. Moreover there was an underlying periodicity in growth with four minima since 1961, resembling the period cycle of galactic cosmic radiation."[1] They postulate that what might explain this correlation could be the tendency of galactic cosmic radiation to produce cloud condensation nuclei, which in turn increases the diffuse component of solar radiation, and thus increases the photosynthesis of the forest canopy. Diffuse radiation penetrates the canopy more than direct sunlight.

They found no correlation between temperature or precipitation and growth rates. It would seem that CO_2 had nothing to do with the growth rates since it had slowly and steadily increased during the period of growth. One can safely predict that the "warmists" will attack or ignore these results. It is also likely that the "deniers," who have been getting more and more publicity, will cite these results.

It is interesting that the apparent hoax involving the Colorado flight of a helium-filled balloon, supposedly with a 6-year-old boy on board, received world-wide attention in October 2009, while the hoax aspects of global warming have received very little attention. Senator Orinn G. Hatch of Utah did, however, compile a large number of anti-AGW statements by scientists, most of them actually involved with the IPCC. It was reprinted by The Science and Public Policy Institute in their SPPI Reprint Series dated September 18, 2009. The title is "UN Climate Scientists Speak out on Global Warming," selected and edited by Hatch from the Senate Minority Report. It includes comments from 101 individual scientists sorted by backgrounds as follows:

- UN IPCC authors: 9
- UN IPCC scientists: 7
- UN IPCC expert reviewers: 12
- NASA: 10
- Other government scientists: 6
- State climatologists: 9
- Academies of science: 10
- Avowed environmentalists: 4
- Noted scientists: 27
- Other Nobel Prize winners: 3

Hatch, in his introduction, writes that the included statements prove there is not a consensus, even at the UN, on the widely touted IPCC conclusion that "Greenhouse gas forcing has likely caused most of the observed global warming over the last fifty years."[2] Hatch notes that the chapter of the IPCC report making that conclusion was reviewed by only 62 of the 2,500 scientist reviewers of the IPCC reports.

A very detailed report, "Climate Change Reconsidered: 2009 Report of the Nongovernmental Panel on Climate Change (NIPCC)," refutes the IPCC conclusions. A PDF file of this 880-page volume can be downloaded at *http://climatechangereconsidered.org*. Printed copies are also available. It probably won't be a best-seller, but it includes the names of 31,478 scientists who signed a petition circulated to thousands of scientists, with a cover note by Dr. Frederick Seitz, past president of the National Academy of Sciences. In his note, Seitz said:

"The United States is very close to adopting an international agreement that would ration the use of energy and of technologies that depend upon coal, oil, and natural gas and some other organic compounds. The treaty is, in our opinion, based upon flawed ideas. Research data on climate change do not show that human use of hydrocarbons is harmful. To the contrary, there is good evidence that increased atmospheric carbon dioxide is environmentally helpful. The proposed agreement would have very negative effects upon the technology of nations throughout the world, especially those that are currently attempting to lift from poverty and provide opportunities to the over four billion people in technologically underdeveloped countries."[3]

These are strong words indeed.

Here are some comments from the petition:

"We urge the United States government to reject the global warming agreement that was written in Kyoto, Japan, in December, 1997, and any other similar proposals. The proposed limits on greenhouse gases would harm the environment, hinder the advance of science and technology, and damage the health and welfare of mankind.

There is no convincing scientific evidence that human release of carbon dioxide, methane, or other greenhouse gases is causing or will, in the foreseeable future, cause catastrophic heating of the Earth's atmosphere and disruption of the Earth's climate. Moreover, there is substantial scientific evidence that increases in atmospheric carbon dioxide produce many beneficial effects upon the natural plant and animal environments of the Earth."[4]

Obviously, the number of signers far outweighs the 2,500 IPCC people. Importantly, none of the circulation funding came from oil or gas companies or other interested parties. Among the signers were many physicists, chemists, climatologists, and other scientists. It is certain that the widely repeated notion that the matter of climate change is settled, and that the unquestioned consensus of the scientific community is that CO_2 is responsible for a rapidly growing worldwide temperature and must be stopped at all costs, is clearly not true.

A very important and detailed study was published by Dr. Habibulto Abdussamatov, head of the Space Research Laboratory of the Pulkovo Observatory near St. Petersburg, Russia. The observatory was built in 1839, destroyed in World War II, and then rebuilt. For some time, it had the largest telescope in the world. With it, Abdussamatov and his colleagues have done a detailed review of the data we have on sunspots and other activities on the sun. His focus was on solar radiation emission as a function of time, with his study's data on sunspot numbers going back to 1611. His provocative title is "The Sun Defines the Climate." The entire paper can be found at *http://climaterealists.com/index.php?id+4254*. An excerpt reads, "Experts of the United Nations in regular reports publish data said to show that the Earth is approaching a catastrophic global warming, caused by increasing emissions of carbon dioxide to the atmosphere. However, observations of the Sun show that as for the increase in temperature, carbon dioxide is 'not guilty' and as for what lies ahead in the upcoming decades, it is not catastrophic warming, but a global and very prolonged temperature drop."[5]

These are strong words backed up by a great deal of information. A key point of Abdussamatov's research is that it had been thought that the amount of energy emitted by the sun was constant in time. Better measurements have shown this isn't the case and there is a periodicity in the energy output of the sun. It had been known since the middle of the 19th century that the number of sunspots on the surface of the sun varies in an 11-year cycle. An English astronomer, Walter Maunder, in 1893 discovered that from 1645 to 1715, sunspots had been essentially absent: only 50 spots were noted in that period, whereas it would have been normal for 50,000 sunspots to have appeared in that time period. We know now that there have been such minima many times in the past. He also noted that the Maunder Minimum, as this low–sun spot period came to be called, included the coldest dip in temperatures that had been noted for thousands of years.

In 1976, an American astrophysicist, John Eddy, noted that there was a correlation between periods of significant change in the number of sunspots and large changes in the climate in the past millennium. In 1988, a soviet geophysicist, Eugene Borisenko, showed that in each of 18 deep minima of solar activity, there have been periods of global cooling.

There were periods of global warming during periods of high sunspot activity. About every 200 years, there are such minima and maxima. It is this bicentennial variation in climate that is so important, even more so than the 11-year solar sunspot level. The primary factor here is that it has been discovered that the TSI (Total Solar Irradiance) is not, as has been thought, a constant, but rather varies in time in a periodical fashion. According to Dr. Habibulto Abdussamatov, the sun is a variable star, which changes its parameters under short and long cycles. The sun is never found in a steady state of energetic and mechanical equilibrium. He points out that an entirely new instrument, a solar Limbograph, is to be installed in the Russian section of the International Space Station in 2011. It should be able to make the most accurate measurements ever of the radius and energy output of the sun, which will allow for much more precise predictions of climate than are now possible. He also points out that the gradual *growth* of ice caps at the poles has already begun—not the melting that some expected.

NASA successfully launched the Solar Dynamics Observatory (SDO) on February 23, 2010. It is planning on obtaining much-improved observations of the sun with a new instrument named EVE—"The Extreme Ultraviolet Variability Experiment"—onboard the SDO. Though the sun appears from Earth to be a consistent, quite placid surface, in extreme ultraviolet frequencies, it is a seething cauldron of storms and prominences, sunspots and faculae.

Two important events took place near the end of 2009. One involved the determination by careful observation by meteorologists that 90 percent of the 1,100 official surface-measuring thermometers in the United States did *not* meet official regulations of their locations not being near sources of heat. This bias was almost always found to raise the apparent temperatures. The second problem was a hacking of the files of the Climate Research Unit (CRU) at East Anglia University and the subsequent release of more than 1,000 e-mails. This has been described as an enormous scandal because the e-mails provide clear evidence of bias, deception, misrepresentation, and suppression of dissent by the CRU. Chairman Jones resigned at the beginning of December, as an outside committee will review the situation. Major media outlets, such as the *New York Times*, the *Toronto Globe and Mail, McLeans* magazine, and even CBC Radio's "As it Happens" program, who had previously gone along with the warmists, have paid attention to the new information. How much impact these revelations had on the looming Copenhagen International Climate Conference is not known. In early 2010, the IPCC lost more public confidence when it was discovered that its claim that large glaciers in the Himalaya Mountains would melt within a few decades was false.

For more information on global warming, read Brian Sussman's very important book, *Climategate: A Veteran Meteorologist Exposes the Global Warming Scam*, published by World Net Daily in April 2010.

FRONTIERS OF SCIENCE

Frontier sciences investigating topics such as extraterrestrial visitation and paranormal phenomena have not gained wide acceptance by mainstream scientists, although they have been studied by open-minded academics for decades. Throughout history it has been difficult, if not impossible, to promote the acceptance of new discoveries. The reasons are multifaceted, but often involve arrogance, egos, politics, greed, and resistance to change. Today's professional "skeptics" often adhere to an almost theistic belief in "science," marked by cynicism and the manipulation of data to fit their personal beliefs. Many plead for scientific scrutiny but are often, in reality, scientifically naive writers. Mainstream scientists, the media, and the general public are often deceived by these skeptics' misinformation. They target frontier sciences, such as psi phenomena and UFOs, and work to suppress scientific exploration in these and other emerging sciences.

Historically, when new paradigms have threatened existing dogma, those who clung to archaic ideology have worked to suppress emerging scientific ideas. Today it is apparent that similar forces are engaging in ad hominem attacks against frontier scientists, disseminating fabrications and misrepresenting factual information. Examples of this type of cynicism are abundant on the Internet. One can only wonder how many groundbreaking discoveries are being suppressed as a result of false propaganda.

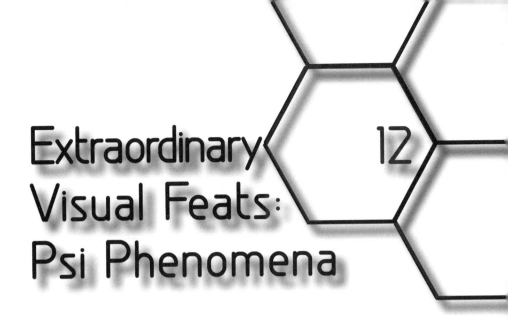

Extraordinary Visual Feats: Psi Phenomena

12

Our popular culture's perception of psychic phenomena conjures up images of "seers" peering into crystal balls, trance mediums channeling messages from dead family members, tarot card and palm readers, spoon benders, house hauntings, and demon possession. Despite its slightly off-kilter public image, a 2005 Gallup poll revealed that 75 percent of the American public believes that at least one paranormal phenomenon is real. Skeptics attribute this belief to a paucity of scientific knowledge due to the failure of our educational system. However, a 1979 survey of 1,100 American college science professors produced surprising results: 55 percent of natural scientists and 66 percent of social scientists believed that extrasensory perception is either an established fact or a likely possibility. That figure jumped to 75 percent when academics in the arts, humanities, and education were surveyed. However, psychologists cast a dissenting vote, indicating that a full 34 percent believe that ESP is impossible, whereas only 2 percent of the respondents expressed the same sentiment.

There is a significant reduction in the percentage of believers in psychic phenomena when politics figures into the equation. A 1992 survey of the 2,000-member National Academy of Sciences, a private organization of scientists and engineers located in Washington, D.C., and an official advisory board to the federal government, found that 75 percent of its members did not believe in the reality of psychic phenomena.

Most people have either experienced what they perceive as psychic phenomena or accept it as real based upon a massive collection of anecdotal evidence. For example, during a vacation in Bermuda, "Harry Milton" experienced a disturbing nightmare in which his wife sustained life-threatening injuries in a motor scooter accident. On the final day of their vacation, she expressed a desire to tour the island via motor scooters, but he strenuously protested the idea due to his disturbing dream. She refused to concede to his superstitious nature and leased a motor scooter. During a trial training exercise, her bike malfunctioned and she was thrown over the handlebars, catapulting head-first into a wall. The C-1 cervical fracture severed her spinal cord and rendered her quadriplegic for the remainder of her life. Harry is haunted by his precognitive dream and his failed attempt to intervene on his wife's behalf.

Alex Tannous Research Library at the Rhine Research Center, Institute of Parapsychology, Durham, N.C. Photo by Kathleen Marden.

"Dave Walters" was skeptical about psychic phenomena, but in 1977 he met Alex Tanous, a nationally known psychic, during a social gathering at his relative's home. Alex approached Dave and asked to speak with him privately. In a quiet room, Alex informed Dave that he could detect a mass in his left abdominal region and advised him to follow up with a competent medical professional. A healthy 33-year-old and somewhat bemused skeptic, Dave made the decision not to heed the medical initiative's advice. Three years later, when he and his wife were expecting their second child, Dave sought the services of a chiropractor to relieve unremitting back pain. X-rays revealed the presence of a large mass on his left kidney. Laboratory tests confirmed renal cell carcinoma. Dave spent the next 14 years in a protracted battle for his life. In 1995, he succumbed to metastatic cancer.

These are examples of anecdotal evidence and are not generally accepted as scientific proof of psi phenomena by the scientific community. The familiar statement "extraordinary claims require extraordinary evidence" has become the mantra of academic scientists. The higher standard to which they subscribe requires substantial statistical evidence from reliable and replicable experimental studies.

A small yet influential group of vocal arch-skeptics claims 100 years of research has failed to produce convincing evidence for psi phenomena. Parapsychologists disagree, pointing to hundreds of flawless scientific studies that produced statistically significant evidence for some psi phenomena. Each group argues that its opponent is guilty of confirmation bias—the tendency to focus upon the evidence that supports one's preconceived beliefs and filters out evidence to the contrary. Therefore, parapsychologists are accused of selectively reporting evidence that's in favor of psi phenomena, while arch-skeptics reject statistically significant experimental replications as flawed.

One example of confirmation bias is evident in the experimental study of Natasha Demkina, a young Russia medical intuitive who claimed she could "see" the full structure of the human body, including how internal organs are positioned and how they function, almost as if she had x-ray eyes. In January 2004, *The Sun*, a British tabloid, funded an experiment to test her claim by asking her to do "cold readings" on two experimental subjects. She reportedly demonstrated a genuine ability for intuitive medical diagnosis by accurately identifying numerous fractures, nerve damage, metal pins, and a titanium plate in a test subject. In the second test, she reportedly misdiagnosed diseased organs in an allegedly healthy subject. Natasha protested that she is able to "see" the early stages of disease before a medical diagnosis is possible. No follow-up medical tests to confirm or refute her claim have been publicized.

Parapsychologists have reported that psychic ability is variable within humans and therefore inconsistent in experimental subjects. We know that this is true for all forms of human performance. Even the best psychics are not 100 percent accurate; all people function better on some days than on others, due to variables such as fatigue, health, the environment, our emotions, weather conditions, and even the phase of the moon. Therefore, extrasensory perception is not consistent from moment to moment or day to day. It is paranormal, not supernatural.

The Discovery Channel produced a television program in 2004 allegedly designed to test Natasha's paranormal ability in an objective and unbiased manner. By all appearances, however, the experiment was designed to increase Natasha's chances of failure. Thus a confirmation bias was already in place when two psychologists and a science and medical journalist, known for debunking the paranormal, designed and implemented the preliminary tests. All three were members of the Committee for the Scientific Investigation of Claims of the Paranormal (currently the Committee for Skeptical Inquiry), a well-known private debunking organization, and its (now defunct) affiliate the Commission for Scientific Medicine and Mental Health. One of its members, Ray Hyman, a professor emeritus at the University of Oregon, has a long history as an arch-skeptic of psi phenomena. He earned his doctorate in psychology at Johns Hopkins University and taught at Harvard for several years. A noted expert in research design and statistical methods, he was capable of designing a high-quality experiment to test Natasha's intuitive ability. Throughout his career, however, he developed a notorious reputation for attributing successful psi experiments to design flaws, bias, or cheating, despite evidence to the contrary. Another member, Richard Wiseman, is a professor in the Public Understanding of Psychology department and head of a research unit at the University of Hertfordshire, U.K. He is a former professional magician who wrote his doctoral dissertation on the psychology of deception and is a well-known debunker of paranormal phenomena. He has never found favorable evidence for psi phenomena. The third committee member, Andrew A. Skolnick, is an award-winning science journalist and photographer known for debunking the paranormal, and the former director of CSMMH.

For their study, Natasha traveled from Russia to New York in May 2004 to participate in what she thought was an objective scientific experiment by some of the world's top scientists. She had naively signed a contract that awarded full control to the experimenters, without expert scientific advice or legal council regarding variables in the experimental design that would interfere with her usual performance. The experimental protocol to which Natasha and her mother agreed exceeded the internationally accepted statistical standard for success in psychology tests—the probability of 1 in 20—indicating bias against a successful finding. The teenager and her uninformed mother agreed that her results would be judged "consistent with chance guessing" if she matched fewer than five of the seven target medical conditions rather than the accepted value of four.[1]

Prior to the controlled experiment, Natasha was instructed to do a visual scan on several people who had reported specific ailments to the scientists. She scanned each of her subjects and reported her impressions regarding the state of their internal organs. Although five out of six were impressed by her ability to accurately diagnose their medical conditions, the experimental team did not concur with their subjective reports. The experimenters noted that she had identified only one of their previously stated primary conditions. Although she had correctly identified secondary medical conditions, this anecdotal evidence was not acceptable to the experimental team.

The scientifically controlled part of the experiment required Natasha to perform in a manner that would not replicate her previous successful experiments in diagnosing diseased organs. She was required instead to diagnose conditions that some of the test subjects no longer exhibited. In one case, two of the subjects had undergone appendectomies, but only one represented a target hit. Furthermore, whereas she usually saw only one patient at a time, she was asked to intuitively diagnose an entire group of six volunteers and one control subject, who presented no diagnosed medical condition. The experimenters presented her with six test cards, each listing a targeted medical condition in English and Russian. The target conditions included an apical resection of the left lung, a surgically removed appendix (actually two), a surgically removed section of the lower esophagus, a cranial metal plate, post-surgical metal staples in the chest area, and an artificial hip joint.[2]

Although under normal conditions Natasha allegedly diagnosed medical conditions in minutes, due to the unfavorable requirements in the experimental design, it took her four hours to complete the test. In the end, she had successfully identified the target conditions in four of the seven test subjects. By international statisticians' standards she passed the test. But the experimental team set a higher bar for Natasha. Citing the fact that she had not attained their five required hits, the team announced that she had failed the test.

A statistical analysis of Natasha's performance reveals that she performed at 50 times the chance probability rating by successfully identifying four out of the seven targets. By international standards she clearly demonstrated an uncanny ability to intuitively diagnose the target medical conditions. However, the experimental design, methodology, and statistical standard set by the experimental team resulted in a confirmatory bias against Natasha.

An international debate ensued when scientists accused the experimental team of foul play. Nobel Prize–winning physicist and director of the University of Cambridge's Mind Matter Unification Project, Brian Josephson, PhD, criticized the experimental team's methodology and questioned their motives. He stated, "A statistically very significant result was obtained in the quantitative part of the investigation, but the experimenters concealed the fact with their talk of failing the test. The investigation claims to be science but fails almost every test of good scientific practice."[3]

Richard Wiseman defended the team's decision to raise the bar above international statistical standards in this statement: "I don't see how you could argue there's anything wrong with having to get five out of seven when she agrees with the target in advance."[4]

Keith Reynolds, a professor emeritus of applied statistics at the University of Greenwich, commented that Wiseman's statement "demonstrates a complete lack of understanding of how experimental data should be interpreted statistically.... The experiment, as designed, had high chances of failing to detect important effects."[5]

Victor Zammit, an attorney, argued that by the International Covenants on Human Rights and Civil Law, "Every person has a fundamental legal right to have his/her reputation protected." But according to Zammit, "Close analysis of the Discovery Channel program shows that throughout the documentary continuous 'illegalities' and negativity were being heaped upon Natasha."[6] He questioned Richard Wiseman's ability as a closed-minded skeptic to produce impartial results, particularly because he has a reputation for changing protocol, without notice, and because he and his team have never found in favor of psi.

Natasha cried foul play when she complained that on the day of the experiment she was subjected to an unusually hostile environment. Her usual interpreter was replaced by one chosen by CSICOP, she was suffering from jet-lag, the experimental design forced her to use an unfamiliar methodology which reduced her ability to demonstrate her gift, and she had been pressured by the experimental team to "hurry up" (a well-known method of reducing psychic ability). Further, she stated scar tissue had interfered with her ability to see the esophageal target area on one test subject, that two test subjects had undergone the same abdominal surgery, but only one of them was the target subject. (In subsequent experimental tests at Tokyo University in Japan, Natasha correctly diagnosed the primary medical conditions in each of seven subjects, and also in a dog at a veterinary clinic.) The Natasha Demkina experiment demonstrates the failure of biased experimenters to pursue scientific truth due their own preconceptions that psi is impossible.

As it turns out, parapsychologists have, in the past one hundred years, worked diligently to reduce design flaws and confirmation bias by designing empirical studies that defy even the skeptics' criticisms. Despite strong resistance from the scientific establishment, statistical evidence from hundreds of experiments demonstrates psi phenomena is real and cannot be explained away as chance or as resulting from flawed methodology or confirmation bias.

Parapsychology laboratories at universities and private institutes moved from the observation of anecdotal evidence, common in the late 19th century and the first decade of the 20th century, to standardized research procedures. In 1917, John E. Coover, a psychologist at Stanford University, used a deck of 40 regular playing cards to conduct 10,000 individual telepathy trials. Test subjects, including "senders" and "receivers"

J.G. Pratt records the results of a basic Zener Card ESP test at Duke Parapsychology Lab. Courtesy of the Rhine Research Center Institute for Parapsychology, Durham, N.C.

were separated in adjoining rooms. The object of the experiment was for the "receiver" to correctly identify the playing card randomly selected by the "sender" at a rate greater than chance. The odds against chance in the successful trials were recorded at a rate of 160 to 1.[7]

In the early 1930s, psychologist Karl Zener, an associate of William McDougall, FRS. and J.B. Rhine, at Duke University's psychology department, developed a special pack of cards for testing ESP that, under controlled conditions, would measure the psi performance of test subjects. The deck of 25 cards consisted of five groups of five symbols: a square, a circle, a star, a triangle, and wavy lines. The "sender "shuffled the deck and selected the top card. This information was then "sent" telepathically to the "receiver" in another location who would attempt to identify it. Experiments from the late 1920s to 1965 resulted in performance scores significantly higher than chance for both telepathy (thought transference of the target information) and clairvoyance (double blinded so that no one knew the target). However, because variables might have skewed the results, improved experimental designs were needed if experimenters were to provide convincing evidence of psi.

In the early 1970s, three independent researchers simultaneously developed a sensory deprivation technique to produce an altered state of consciousness under which psychic abilities might be enhanced. Charles Honorton at the Division of Parapsychology and Psychophysics at Maimonides in Brooklyn, New York; Adrian Parker, associate professor in psychology at the University of Gothenburg, Sweden, then at the University of Edinburgh; and William Braud, an experimental psychologist at the University of Houston; developed "ganzfeld" (German for "whole field") telepathy experiments that produced acceptable scientific evidence for psi phenomena. Researchers and skeptics had jointly agreed to specific guidelines for experimental methodology and measurement, thereby reducing design flaws. The ganzfeld was deemed superior to earlier experiments because it clearly delineated the experimenter, the "sender," and the "receiver."

In a typical ganzfeld experiment, the receiver is isolated in a soundproof room wearing translucent ping-pong-ball halves as eye shields through which red light is projected as a stimulus. This reduces sensory input with regard to the room's physical environment. The receiver listens to a relaxation tape followed by "white noise" (a sound similar to soothing radio static) through headphones. The sender is isolated in a separate soundproof room. A closed-circuit video system repeatedly presents a static visual target image, such as a nature scene, over a monitor, and the sender attempts to mentally transmit the target image to the receiver. For the next 30 minutes, the receiver gives a constant flow of verbal reports regarding mental imagery and thoughts that enter his/her mind. The experimenter (sequestered in a third location), and the receiver are unaware of the target. At the end of the 30-minute period, the receiver is presented with four visual stimuli (for example, four unrelated nature scenes). One contains the target image. The receiver is asked to rate the degree to which each image matches the mental imagery and thoughts experienced during the experiment. If the receiver assigns the highest rating to the target image, it is recorded as a hit. All others are recorded as misses. The hit rate by chance alone is 25 percent.

In 1982, Charles Honorton presented a paper at the annual Parapsychological Association convention. His analysis, the first meta-analysis of all known ganzfeld experiments to date, provided significant evidence for the existence of psi. Arch-skeptic Ray Hyman (a Demkina experimenter) disagreed and conducted an independent analysis of the same studies. This led to two independent meta-analyses by Ray Hyman and Charles Honorton to evaluate the ganzfeld experimental results. Hyman examined 42 psi ganzfeld studies in 34 reports from 1974 to 1981 in which he found some evidence of methodological flaws and experimenter bias. Honorton agreed with some of Hyman's criticisms and made statistical adjustments to correct the bias, but disagreed with Hyman on other points. Eventually they both agreed that the results could not be attributed to either selective reporting or chance.

Independent raters, including two statisticians and two psychologists, concurred with Honorton's conclusion that flaw variables did not significantly affect the study outcome. Thus Honorton's results favoring the existence of psi gained independent support. The odds against chance were 10 billion to one. In 1986, Hyman and Honorton issued the following joint statement regarding the 1985 meta-analysis: "We agree that there is an overall significant effect in this data base that cannot reasonably be explained by selective reporting or multiple analysis. We continue to differ over the degree to which the effect constitutes evidence for psi, but we agree that the final verdict awaits the outcome of future experiments conducted by a broader range of investigators and according to more stringent standards."[8]

In 1987, the National Research Council of the National Academy of Sciences issued the following statement: "The committee finds no scientific justification from research conducted over a period of 130 years for the existence of parapsychological phenomena."[9] The NRC's evaluation was based upon the meta-analysis conducted by arch-skeptic Ray Hyman (the chairman of the NRC's subcommittee on parapsychology), and did not include an independent examination of the ganzfeld database. Accusations of bias led to an investigation, which found that the committee had asked Robert Rosenthal of Harvard University to withdraw his conclusions because they were favorable to psi. The investigation committee concluded that parapsychology deserves "a fairer hearing across a broader spectrum of the scientific community so that emotionality does not impede the objective assessment of experimental results."[10]

Honorton and his colleagues initiated a new series of ganzfeld experiments designed to eliminate the methodological problems identified in the Hyman/Honorton meta-analyses. Well-known critics of parapsychology served as consultants, including two mentalists who ascertained that the experimental protocols were not subject to sensory leakage or intentional deception. All experimental data had to be reported, thus eliminating the selective reporting of only experiments that favored psi. The experiments continued until 1989, when a loss of funding forced the laboratory to close.

In all, 240 receivers took part in 354 experimental sessions. They used the same basic procedure we have described, but ascertained that the sender and receiver were isolated in sound proof, electrically shielded rooms. A computer presented a random target image to the sender. Half of the experiments presented static targets such as photographs, art prints, and such, and the other half presented dynamic targets such as movie clips, cartoons, documentary clips, and the like. While the sender watched the random images on a screen, the receiver communicated his/her impressions about the target image, which was also tape-recorded. At the end of the 30-minute ganzfeld experiment, a TV monitor in the receiver's isolated room presented one target clip and three non-target clips to the receiver. The receiver used a computer game paddle to make ratings on a 40-point scale presented on the monitor. The computer then wrote the results onto a file on a floppy disk, thus eliminating the possibility of bias. When this was complete, the computer revealed the target information to the receiver and the experimenter.

The autoganzfeld meta-analysis delivered statistically significant results in support of psi, which were consistent with those in the earlier Hyman/Honorton meta-analysis. The results produced odds against chance of 45,000 to 1. In 1991, Ray Hyman wrote, "Honorton's experiments have produced intriguing results. If independent laboratories can produce similar results with the same relationships and with the same attention to rigorous methodology, then parapsychology may indeed have finally captured its elusive quarry."[11]

The autoganzfeld studies revealed that dynamic targets (movie clips) produced a higher hit rate than static targets (pictures). Believers in psi and those who practice meditation scored higher than those who did not. Extroverts scored higher than introverts. Test subjects from the Juilliard School in New York City were strong performers in the psi experiments, suggesting a relationship between ESP and creativity or artistic ability. Musicians achieved significantly higher results. Also, experienced "receivers" had a significantly higher hit rate than novices, even though each experiment presented new target material. Honorton and his associates advised future researchers to select "receivers" who exhibit characteristics consistent with psi success rather than through random selection. Additionally, they advised researchers to create a warm social climate conducive to psi success, as opposed to a negative or stressful social climate that correlates with psi failure.

The autoganzfeld experiments received considerable attention when, in 1994, the mainstream journal *Psychological Bulletin* published the results of a new meta-analysis by Daryl Bem, a psychology professor at Cornell University, and Charles Honorton.

Convinced they had successfully demonstrated the replicability of the ganzfeld visual telepathy experiments, some experimenters modified the procedure. The new ganzfeld experimental design tested whether senders could telepathically transmit musical targets to receivers. The musical target tests provided evidence that auditory signals could not be transmitted telepathically.

Psychologist Julie Milton, of the University of Edinburgh, and skeptic Richard Wiseman (of the Demkina experiment) undertook the meta-analysis of 30 new ganzfeld studies, including the failed auditory target experiments, which began after 1987 and were published by early 1997. Their meta-analysis failed to confirm psi.

Proponents of psi countered that the tests that used non-standard ganzfeld procedures (auditory telepathy) deviated significantly from standard ganzfeld experimental design and should not have been included in the Milton-Wiseman meta-analysis. A joint effort between Cornell University and the Rhine Institute for Parapsychology followed, which added 10 studies that had been completed but not published during the Milton-Wiseman meta-analysis to the M/W database. An Internet debate ensued in which Julie Milton stated that when replications after the Milton-Wiseman cutoff date are added to the meta-analysis, the studies do achieve statistical significance in favor of psi.

Independent replication of controlled experiments performed thousands of times by researchers around the world has demonstrated statistical evidence in support of psi phenomena. This extraordinary claim has been supported by extraordinary evidence. Although the hit rate in experiments averages 32 percent, among telepathic people, such as psychics, it is 65 percent—pretty amazing despite the strong social prohibition against it by Western science.

Dean Radin, PhD, director of the Consciousness Research Laboratory at the University of Nevada, summarized the ganzfeld meta-analyses as follows: "From 1974 to 1997, some 2,549 ganzfeld sessions were reported in at least forty publications by researchers around the world.... A 1985 meta-analysis established an estimate of the established hit rate.... A six-year replication was conducted that satisfied the skeptics calls for improved procedures...that showed the same successful results...psi effects do occur in the ganzfeld."[12]

In 2004, the Institute of Noetic Sciences conducted a technically updated Ganzfeld experiment in 13 pairs of volunteers. The receiver relaxed in an electromagnetically shielded sound-proof room while the sender, in a remote dimly lit room, was randomly stimulated by a video image of the receiver. The receiver remained silent in this test, but an EEG recorded the nervous system responses from both individuals. The experiment produced positive results of EEG correlations as evidence of psi.

Another meta-analysis by two skeptics in 2005 resulted in a 32-percent statistically significant result. Feeling compelled to debunk psi, they used an ad hoc model of how psi might work and refuted it on that flimsy basis. Today, functional MRI tests are recording evidence of psi during modified ganzfeld experiments, which reveal positive effects on the visual cortex of "receivers" in random stimuli presentations.

Despite 130 years of scientific evidence in support of psi effects, the controversy rages on. Could it be that all psi researchers produce flawed experiments? Or is it more likely that skeptical scientists refuse to acknowledge evidence of psi phenomena? When replicable scientific evidence conflicts with current paradigms of natural science, conservative scientists are reluctance to change their a priori assumptions about physical reality. They adhere to the assumption that psi phenomena are impossible. It follows that arch-skeptics are more motivated than mainstream researchers to deny the existence of psi because they refuse to accept the idea that psi exists. As has been demonstrated in the meta-analyses, controlled experimental studies have clearly resulted in statistically significant evidence of telepathy. However, arch-skeptics have consistently denied the empirical evidence, often passing judgment without inquiry, discrediting rather than investigating, ridiculing through ad hominem attacks and charges of fraud, misrepresenting the facts, and making unsubstantiated claims.

Charles Honorton attempted to bring about attitude change by engaging a team of skeptics in his design methodology in order to eliminate experimental flaws. When his new, controlled experiments clearly produced statistically significant results, consistent with earlier published data, arch-skeptics would not alter their a priori beliefs. Instead of considering the possibility that our current model of physical reality is incorrect and might be explained by quantum theory, in which several phenomena have already been confirmed experimentally, the arch-skeptics cling to their belief system and label parapsychology a pseudo-science. Sometimes emotionality takes precedence over rationality, particularly when we wish to hide our heads in the sand to protect our firmly held beliefs. One can only wonder how long it will be before arch-skeptics accept scientific findings for psi phenomena.

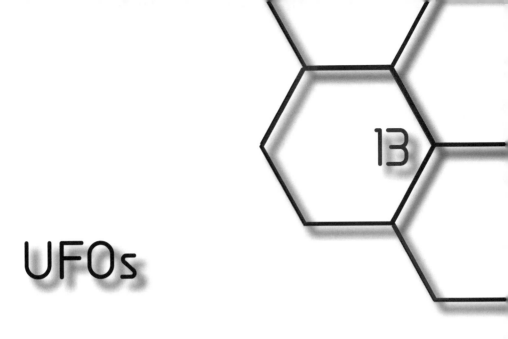

UFOs

The impossibilists have had a field day with UFOs for more than 60 years. There has certainly been no shortage of strong, negative proclamations from debunking groups and individuals who refuse to examine the evidence. These proclamations include the false assumption that there is no convincing evidence, only anecdotal data, to support the notion that some UFOs are of extraterrestrial origin; that eyewitness testimony cannot be trusted; that it would be impossible to get to Earth from another star or another galaxy; that the reported maneuvers would be impossible; that governments can't keep secrets; and many other equally foolish statements.

It is truly amazing how often one finds scientists making claims about the absence of evidence relating to UFOs. They reference tabloids, but not refereed scientific journals such as *The SEE Journal* or *Journal of UFO Studies*. Nor do they examine the body of evidence in large-scale scientific studies and doctoral theses, of which there are at least a dozen. All concede that there have been loads of anecdotal reports of so-called UFOs, but insist there is not a shred of physical evidence.

The late Carl Sagan, a prominent astronomer at Cornell and part of the Air Force's Scientific Advisory Board, probably did more than any other scientist to get people interested in the notion of extraterrestrial intelligent life. He appeared on popular television shows, such as Johnny Carson's *Tonight Show*, and received attention from the American media. His TV program *Cosmos* was seen by literally hundreds of millions of people all over the world, and the book version had enormous sales. Yet, while he didn't ignore the question of UFOs, he *did* ignore the evidence. Though he edited the book *UFOs: A Scientific Debate*, which included papers by a number of contributors, his references to the major scientific studies on UFOs are essentially nonexistent in his work. For example, he claimed, "There are interesting sightings that are unreliable and reliable sightings that are uninteresting, but there are no sightings that are reliable and interesting." This is an extraordinary claim considering the scientific evidence to which he had access. It was Carl who often stated that extraordinary claims require extraordinary evidence. Yet, it is apparent that he chose to ignore the extraordinary evidence.

There is certainly data indicating the reliability of interesting sightings. The Battelle Memorial Institute in Columbus, Ohio, an organization that does a considerable amount of highly classified government-funded research and operates a number of federal laboratories, conducted the largest UFO study ever done for the United States Air Force (USAF). The study was under contract to the Foreign Technology Division of the USAF in Dayton, Ohio, and titled "Project Blue Book Special Report No. 14." Its stated purpose was to determine whether or not any of 3,201 UFO reports "represented technological developments not known to this country," and to construct a model of a UFO based upon the data. The USAF issued a widely publicized official press release on October 25, 1955, in which Donald Quarles, the Secretary of the Air Force, stated, "On the basis of this study we believe that no objects such as those popularly described as flying saucers have over-flown the United States. I feel certain that even the unknown three percent could have been explained as conventional phenomena or illusions if more complete observational data had been available."

This claim was a direct contradiction of the findings, and totally false. The Battelle Memorial Institute did a quality (reliability) evaluation of each of the 3,201 cases and listed each report in one of several categories, including astronomical, balloon, and so on, as well as "unknown" and "insufficient information." No report could be listed as unknown unless all four final report evaluators agreed that it was a true unknown. Any other designation required only two votes.

Quarles, and later Sagan and a few debunkers, claimed that a meager 3 percent of the unknown sightings could not be explained. However, an analysis of the data reveals that a statistically significant 21.5 percent, not 3 percent, were listed in the unknown category. The 9.3 percent for which there was insufficient information are listed separately from the unknowns. Furthermore, the higher the quality of the sighting, the *more* likely it was to be listed as unknown rather than in the category of insufficient information. This is exactly the opposite of what both Quarles and Sagan claim. A full 35.1 percent of the "excellent" sightings were unknowns, while only 18.3 percent of the "poor" cases were so designated. There is nothing surprising here. Obviously one would expect that the better the quality of the case, based upon the reliability and experience of the observers, the duration of observation, the distance from the witness, and so on, the more likelihood that it would be listed as an unknown, if it truly represented a technological development not known in the United States.

In another scientific study, presented in the "Proceedings of the Symposium on Unidentified Flying Objects" held by the Committee on Science and Astronautics of the United States Congress on July 29, 1968, James E. McDonald, associate director of the Institute of Atmospheric Physics and a full professor of physics at the University of Arizona, presented a 71-page report on his extensive research on UFO data. He had talked to more than 500 witnesses, and many of the best cases are included among the 41 cases he presented in his report "Statement on Unidentified Flying Objects."

Carl Sagan made numerous attacks on UFO reality, but failed to discuss the positive findings in any of the large-scale scientific studies in his book *The Demon Haunted World: Science as a Candle in the Dark*. It is uncanny that Carl, a respected scientist, would reference articles

Dr. James E. McDonald, physicist. Courtesy of the University of Arizona.

from *Weekly World News*, a tabloid, but not the Proceedings of the Congressional Hearings (at which he was one of six scientists that testified in person), or any of the other high-quality scientific inquiries into the UFO evidence.

Had Carl Sagan investigated, he could have also mentioned the significant findings of the University of Colorado Study on UFOs. Edward U. Condon, a physicist and theoretician at the University of Colorado, headed a research project funded by the U.S. Air Force to investigate a number of cases and to determine the benefits of closing Project Blue Book, its official UFO investigative arm. In the end, a significant number of cases examined by the Condon committee remained unexplained. Observers protested that the number should have been higher and pointed to the fact that Condon avoided cases that warranted serious attention, and that he clearly presented a negative tone in his statements to the press. A few of the unexplained cases he avoided were so puzzling to the reviewers they had difficulty denying the reality of unconventional flying objects.

Surely Sagan read the infamous, widely publicized "trick" memo, removed from Condon's assistant project director's files, that stated, "The trick would be, I think, to describe the project, so that to the public, it would appear to be a totally objective study but, to the scientific community, would present the image of a group of nonbelievers trying their best to be objective, but having an almost zero expectation of finding a saucer."[1]

Despite the official statement made by Edward Condon that the committee found no evidence to justify a belief that extraterrestrial visitors have penetrated our skies, the special UFO subcommittee of the American Institute of Aeronautics and Astronautics found that 30 percent of the 117 cases studied in detail could not be identified. As we have seen throughout this book, the truth is in the body of evidence, not in the negative but politically expedient official statement.

Carl Sagan didn't even pay lip service to the 1972 book _The UFO Experience: A Scientific Inquiry,_ even though it was written by the late J. Allen Hynek, chairman of the astronomy department at Northwestern University and the Air Force's Scientific Consultant to Project Blue Book for more than 20 years. It would have made an excellent source. Might we postulate based upon the absence of evidence in Sagan's reports that he chose to conceal or ignore the good-quality evidence while disseminating false propaganda to the American public? It appears that this could be the case.

Dr. J. Allen Hynek. Courtesy of Northwestern University.

Sagan and other debunkers have lamented, "there is not one example of physical evidence that sustains even the most casual scrutiny."[2] This statement is completely false. It ignores physical evidence, such as radar-visual cases including film of radar screens and tapes of voice communications with pilots, of which there are many. For example, the famous RB-47 case of July 17, 1957, involved a highly trained Air Force crew of six operating a sophisticated reconnaissance aircraft in a sighting lasting 1.5 hours and covering 700 miles. The UFO was observed visually and by the plane radar. Signal-receiving equipment also registered a signal coming from the object. In addition, ground radar observed both the unknown and the RB-47. James McDonald interviewed all the crew members, and the case is also described in an article in _Astronautics and Aeronautics_ magazine. Isn't it strange that measurements of automobile speeds by police officers operating their hand-held radar units are accepted in court as evidence for speeding, but multiple-witness radar visual sightings involving prolonged observations by highly trained people using sophisticated instrumentation are casually ignored?

For 40 years, Ted Phillips of Missouri, a protégé of the late Dr. Hynek, has been collecting reports of physical trace cases from all over the world. To date, the number of such cases exceeds 3,800 from 90 countries. About one-sixth involve reports of beings associated with craft seen on or near the ground and usually observed to land or lift off by more than one witness. Traces have been photographed, and measurements have been made of the composition of the soil in the trace versus nearby soil. Often, the affected soil will no longer support plant growth, in contrast to the control samples, and will also not absorb moisture. The findings are truly anomalous and not the result of fungus, lightning strikes, or the like.

It is often claimed that there is only anecdotal data available about UFOs—for example, somebody reported seeing a vague light in the sky. This is more pseudoscience without foundation in reality. According to Webster's Dictionary, "anecdote" is "a short story narrating a detached incident or fact of an interesting nature; a biographical incident; a single passage of private life."[3] How can one ignore the myriad of detailed investigations of important sightings, with evidence that has been published?

Often debunkers will claim that we all know people can be mistaken, so eyewitness testimony is not to be trusted. Of course they can be mistaken, but the legal system accepts their testimony for review in court. This claim totally ignores the simple fact that the reason most sightings *can* be explained is that eyewitness testimony is reliable. For example, meteorites have been located on the ground as a result of testimony provided by witnesses to the brief trajectory of the meteor. Obviously, the testimony was reliable.

Similarly, when a bright, star-like object is reported, and the astronomical tables are consulted, we often find that Venus had the precise appearance and location of what was described. In that case, the observation was accurate, but the interpretation was wrong. If, however, multiple reliable witnesses report the 10-minute sighting of a large, silent, internally lighted, disk-shaped craft hovering less than 200 feet above the ground, which suddenly shoots vertically into the atmosphere and disappears within seconds, is it safe to assume that they experienced a perceptual aberration? Can we accept only reports that can be explained as mundane events and reject all reports that cannot be explained? To do so would be unscientific.

There were several cases in Southern California in which witnesses described a high-speed bright object moving down the coast around dusk. Calls to Vandenberg Air Force Base up the coast brought forth the fact that Vandenberg AFB had launched a rocket down the coast at just the right time so that it was still in sunlight. The witnesses' observations were accurate, but the interpretation that it was a UFO was wrong. So, either witnesses are doing a reasonably good job of observing or they aren't. To suggest that they are accurate only when what is observed turns out to be conventional is absurd.

One of the most common objections to the idea of alien visitations is based on the notion that travel to other solar systems is impossible and that the details of the sightings don't matter. These claims often come from astronomers and others who know nothing about space travel, but still insist it would take too much energy, would take too long, would violate the laws of physics, and so on. This evaluation should be made by people who possess expertise in space travel, such as scientists who work in the field of aeronautics, astronautics, and deep space travel, as well as theoretical physicists.

As noted in Chapter 1 about flight in the atmosphere and in Chapter 2 about space travel, the self-styled experts have almost always been wrong, because they make totally inappropriate assumptions and have not consulted the relevant scientific literature. In addition, they often set up straw men, such as assuming one must go to Andromeda (more than two million light-years away) or across the Milky Way galaxy (about 100,000 light-years). Would it seem more rational to focus on nearby solar systems, rather than distant ones? There are well over 1,000 stars within 55 light-years, of which at least 5 percent are similar to the sun and might be expected to have planets. Even with our still very crude planet-seeking systems, we have already discovered more than 453 exoplanets. The number will surely increase when the Kepler Space Observatory is fully operational and other advanced space observatories are launched.

Another standard ploy from the debunkers is to invoke Einstein's limit of the speed of light. Normally, they neglect to mention that time slows down for things moving close to the speed of light and that it only takes one year at 1 G acceleration to get close to that velocity, and that both Mother Nature and accelerator physicists (for example using the Large Hadron Collider) provide particles moving at more than 99.999 percent of the speed of light. In addition, they neglect the all-important fact that Mother Nature provides much of the energy for our deep space trips, not the launchers. For example, the Cassini spacecraft was sent around Venus, then Earth, and then Jupiter to receive "free" gravitational assists on its way to Saturn. Dr. Campbell, as noted in Chapter 2, was wrong by a factor of 300,000,000 in his calculation of the required initial launch weight of a chemical rocket able to get a man to the moon and back. Furthermore, the UFO debunkers almost always seem to be unaware of the benefits of using nuclear energy, which provides so much more energy per pound than do chemical propellants. In the case of fusion, which provides the energy of all the stars (and H-bombs), it has been demonstrated many times that it can supply 10 million times as much energy per pound as do chemical reactions.

Yes, the Saturn V rocket that launched the Apollo missions was a behemoth, because chemical rockets are very much less efficient than nuclear systems, but a number of nuclear fission rockets have already been tested, including, for example, the 7-foot-diameter Phoebus 2-B reactor operated by Los Alamos at a power level of 4,400 megawatts back in 1969. That is twice the power of Grand Coulee Dam. Nuclear-powered aircraft carriers operate for 18 years without refueling. That saves an enormous amount of petroleum fuel. We Earthlings figured out in 1938 that the stars produce their energy not by burning gas or even by nuclear fission, but by nuclear fusion. In late 1952, America tested a fusion bomb producing as much energy as exploding 10 million tons of TNT. There have been many papers on the use of nuclear fusion for deep-space rockets exhausting particles having 10 million times the energy per particle as in the Saturn V. In short, there are many ways to produce a great amount of energy to propel spacecraft.

A very common response to reports of flying objects demonstrating great acceleration, right-angle turns, and other non-airplane maneuvers is that they are clearly impossible and would violate the laws of physics. In Chapter 2, we find the claim that when one gets to 9 Gs, one dies. This is nonsense, unless one slams into a wall and isn't wearing seat belts, shoulder harness, and a G suit. Let us be clear: the laws of physics say nothing about how many Gs one can stand. The limits depend upon a number of well-studied factors as well as biology, including the magnitude of the acceleration, the direction with regard to the body, and the duration of acceleration. A trained pilot can perform a tracking task while being accelerated at 14 Gs for two minutes.

The impossibilists have a real problem with possible alien motivation, usually because they seem to be unfamiliar with human motivation. The U.S. military budget for 2009 is about $600 billion. Much of that is spent on a wide variety of surveillance systems operated by such agencies as the National Reconnaissance Office, the National Security Agency, the CIA, DIA, FBI, OSI, ONI, and other intelligence agencies. The purpose is to monitor the activities of potential enemies in order to prevent tragedies like the Pearl Harbor attack of December 7, 1941.

Though we know little about aliens, it would seem reasonable to assume that every civilization would surely be concerned with its own survival and security. Thus, they would have to keep tabs on primitive societies in their neighborhood, but only close tabs on those showing signs of soon being able to bother them. Consider all the technological changes on Earth between, say, 1900 and 1945. There were at least three signs at the end of World War II indicating that soon (on a cosmic time scale) Earthlings, whose primary activity is tribal warfare, would be able to bother extraterrestrials. During the war, 1,700 cities had been destroyed and 50 million Earthlings were killed by their own kind. What came out of the war were nuclear weapons, powerful rockets such as the German V-2, and radar, indicating the beginning of the electronic age.

Isn't it amazing that the only place on Earth in July 1947 where all of this could be studied was southeastern New Mexico? The first A-bomb was exploded at the White Sands Missile Range in July 1945. V-2s captured from the Germans were being tested there postwar, and our best radar was tracking the missile flights. Roswell is also in southeastern New Mexico. There may be a host of other reasons for aliens coming here, but surely self-interest would be one. (Possible alien motivations for visitations are discussed in detail in Stanton Friedman's book *Flying Saucers and Science.*)

Some have objected strongly to the notion of an alien spacecraft having crashed in New Mexico in July, 1947, on the grounds that a craft sophisticated enough to reach Earth from a distant planet could not possibly have crashed once it arrived. In the first place, all indications are that what crashed near Corona, north of Roswell, was small and therefore very likely an Earth excursion module for use in the atmosphere, as opposed to a huge mother ship used to transport the small craft from another solar system. Triangulation by civil engineer Martin Jacek, based upon the testimony of more than 30 witnesses to a huge mother ship seen in the Yukon, indicated the craft was between 0.6 and 1.2 miles long. No report of such huge vehicles crashing is on record, though much smaller ones seem to have crashed near Aztec, New Mexico, in March of 1948, according to the detailed research of Scott and Suzanne Ramsey, and near Varginha, Brazil, in 1996. One must note that two NASA space shuttles, despite their comparative sophistication, have crashed. Pilot error, component failure, and unexpected, unpredictable events have been responsible for many fatal crashes of airborne vehicles. It would appear that aliens are not any more perfect than Earthlings are.

A rather strange objection to UFOs is that if a saucer had crashed with bodies on board, their buddies would have retrieved the bodies. It is claimed that no such effort was conducted, so therefore there was no crash, or, if there was one, it was a remotely controlled unmanned aerial vehicle. This theory is based on two quaint notions:

1   **No effort was made to retrieve.** In fact, there were a number of sightings of UFOs not too far from the New Mexico crash site, and we don't know what they were doing—we certainly don't know if there had been negotiations for the return or preservation of alien bodies.

2   **The United States Marines try their darndest to make sure bodies aren't left behind.** Surely we have no reason to assume that the U.S. Marines and aliens adhere to the same rules. Perhaps as an advanced civilization they do not adhere to Western science's materialistic model. Reports from alleged abductees consistently suggest that

they don't. Maybe aliens consider working beings to be expendable or to be merely containers for souls.

One of the stranger criticisms of the notion of alien visitation is that they can't possibly look anything like us—that is, the way they are described by observers as being humanoid. However, the case has been made that intelligent beings should have their sensory organs at the top, so they can spot enemies more quickly. Two ears can provide binaural hearing, and another doesn't help much. Two eyes provide depth perception. One must be able to manipulate large and small items, therefore requiring something like arms and fingers. Legs would be required to escape from some enemies. In addition, we don't know that we don't have a common ancestry with the visitors. We also don't know that non-humanoid ammonia-breathers, for example, don't go to planets having an ammonia atmosphere. Science requires that we collect and review the data we have, not pine for that which we don't have. It also requires the admission of "I don't know." We should not refuse to investigate topics that deviate from the scientific establishment's point of view.

For reasons unknown, there are professionals who believe they know how alien visitors would behave. The late Isaac Asimov, a very prolific author of both science and science fiction books, has claimed that aliens would either make themselves known or hide from Earthlings, and that, if they do neither, there are no alien visitors! Dr. Asimov provided no basis for this conclusion, and it flies in the face of worldwide reports of unusual vehicles being observed over big cities and small, sometimes abducting Earthlings, sometimes chasing aircraft, and sometimes landing and taking off from the wilds. Surely as creative as he was, he could imagine behavior somewhere between the extremes. Others say, "Why don't they land on the White House lawn?" This presumes they haven't tried, although we know there were many sightings, visual and radar, of UFOs over Washington, D.C., in the summer of 1952. Military jets chased the saucers away.

It should be obvious that alien craft are well aware that they have been spotted by many different radar networks, because it is easy to determine when search radar is locked on. They certainly know we know they are here. They would, these days, also know about our spy satellites monitoring from above. They would undoubtedly know that attempts were made, at least in 1952, to shoot them down per official U.S. Air Force orders.[4] That is hardly a friendly way to respond to visitors. It must also be noted that no request for permission to fly U-2 spy planes over the Soviet Union during the Cold war of the 1950s was made—why would any one expect aliens to ask for permission to monitor Earth's skies and inhabitants? Considering the enormous amount of activity conducted in secrecy by various governments, it is difficult to imagine that we would be told if permission had been requested.

There seem to be a number of people who are convinced that governments couldn't keep a secret such as the recovery of a crashed flying saucer for 62 years. Astronomers and SETI specialists are the loudest voices in this school of thought. Almost invariably, they haven't worked on highly classified programs and have no idea of the vast expenditures of such things, especially those known as "black budget" programs. These items don't show up directly in congressional budgets. The problem is that academics can't conceive of multi-billion-dollar programs off the books. The Manhattan Engineering District, established to develop the atomic bomb during World War II, eventually involved more than 60,000 people and the expenditure of more than $2 billion, which was a great deal of money at the time. Though he was vice president of the United States, Harry Truman was not aware of the program until 13 days after he became president upon the death of President Roosevelt.

The SETI specialists insist that they will keep no secrets. When they receive an alien radio signal, and it has been verified using their protocol, they will tell the world. Such an event would certainly increase their research budgets, but these amounts are very definitely small potatoes compared to those for the development of a variety of spy satellites and high-performance aircraft such as the stealth fighter and the stealth bomber.

Secrets are easy to keep. The keys to doing so are the need-to-know concept and compartmentalization. The only people who have access to the secret information must have an appropriate security clearance and a "need to know" for that particular information. Some people seem to think everybody with a Top Secret clearance has access to all Top Secret materials. This is totally untrue. Others have claimed that all classified material is declassified after a certain period of time, such as 25 years. Again, this is simply not true. Other debunkers take a different tact: discovery of alien life would be the greatest discovery in man's history (hardly, if it only involved a radio signal from a distant planet), and thus would be announced. But consider the implications of saucer visitation: the pilots of a saucer have very advanced flight technology that every nation would very much like to be able to duplicate, because it would provide better weapons-delivery and defense systems. The basic rule is very straightforward: one can't tell one's friends without telling ones enemies. All countries assume they have enemies (Otherwise why the huge defense expenditures?). Underlying this basic concern is the simple fact that everybody in power wants to remain in power, which can be done by keeping certain information a secret.

Yes, some formerly classified documents about UFOs have been released by U.S. government agencies. Audiences immediately start laughing when they are shown blacked-out CIA UFO documents and whited-out NSA UFO documents. Under the Freedom of Information Act, one can obtain 156 pages of the latter and can typically read one or two sentences per page with the rest whited out. No cover-up? Here is one simple sentence that illustrates the problem: USAF General Carroll Bolender stated in October 1969, "Reports of UFOs which could affect national security are made in accordance with JANAP 146 or Air Force Regulation 55-11 and are not part of the Blue Book System."[5] Not the *New York Times*, *Washington Post*, *Time*, nor *Newsweek* have noted this shocking statement. These reports would, of course, be the most important cases.

In summary, there is an enormous amount of information indicating that Earth is being visited by advanced alien spacecraft, despite vociferous claims to the contrary by educated individuals who have not reviewed that evidence. There is no question that substantial information has been withheld and that the media and scientific communities have avoided their responsibilities. The time for both groups to live up to their public responsibilities for seeking the truth is now.

The
Conundrum
of Alien Abduction

The scientific investigation of alien abduction is perhaps the most contentious of all the frontier sciences. Impossibilists would have us believe that there is no evidence to support the hypothesis that some people have been abducted by non-human beings, despite the fact that researchers have been collecting relevant evidence for nearly 50 years. They proclaim that all abductees are delusional, fantasy prone, scientifically naïve, fraudulent, or simply mistaken. Astronomer Carl Sagan's statement, "extraordinary claims require extraordinary evidence," is often repeated by impossibilists. But abduction researcher Budd Hopkins argues, "Shouldn't we be saying instead that an extraordinary phenomenon such as this *demands an extraordinary investigation?*"[1] Unfortunately, few mainstream scientists have the fortitude to risk their careers by investigating this subject cloaked in taboo. History has demonstrated that those who do examine the evidence risk being denied tenure, and those who have tenure risk being forced out of the university. The only institutionally acceptable basis for scientific research is founded upon the a priori belief that "the least likely explanation for unidentified flying objects is the hypothesis of extraterrestrial visitation by intelligent beings."[2]

That quote was made by the National Academy of Sciences in 1969 in its review of the notorious Condon Report, based upon the biased and unfounded conclusions of Edward U. Condon, a physicist and theoretician at the University of Colorado. It is well documented that Condon personally focused upon "crackpot" cases and avoided cases that warranted serious attention, clearly presenting a negative tone in his statements to the press.

At the project's inception in 1966, Robert J. Low, project coordinator, expressed the concern in his infamous "trick" memo: "In order to undertake such a project, one would have to approach it objectively. That is, one has to admit the possibility that such things as UFOs exist. It is not respectable to give serious consideration to such a possibility.... The very act of admitting these possibilities just as possibilities puts us beyond the pale.... I can quite easily imagine, however, that psychologists, sociologists, and psychiatrists might well generate scholarly publications as a result of their investigations of saucer observers."[3]

Is it any wonder that Section 7 of The National Academy of Science's report recommends that UFO reports should be of interest to social scientists? This suggests that project's conclusions and recommendations had been determined before even one evidentiary file was examined. It is no surprise that today it is social scientists—not physical scientists—who receive grants to perform research on alleged close-encounter witnesses and alien abductees in an academic setting. Although there is a massive body of supportive evidence, the analysis is, for the most part, done with scarce private funding.

The primary focus of academic investigations is not upon alien abduction per se, but psychological explanations based upon the a priori belief that alien abduction is highly improbable. This is due in part to the fact that some aspects of the abduction experience deviate so far from Western scientific materialism that the very idea it could be real seems ludicrous to many scientists. Social researchers therefore have viewed alien abduction as a psychological aberration, not a physical experience. Boundary deficit disorder, fantasy-proneness, hallucinations, sleep anomalies, confabulation in hypnosis, false memory syndrome, and cultural mythos have all been offered as psychological explanations for alien abduction. Arch-skeptics have been quite creative in their attempts to explain alien abduction, and have devoted considerable time to attempts to back up conjecture with personality inventories and psychological experiments. Academia seldom examines the scientific evidence that suggests some abductions by extraterrestrial beings could be real, nor does it measure the probability of abduction assessed on a case-by-case basis. That important responsibility falls upon scientific ufologists, most of whom fund their own research and investigation, sometimes with assistance from wealthy philanthropists and civilian UFO groups.

Years of systematic study has indicated that alleged alien abductees exhibit no more psychopathology than the general population. Abductees come from all societal levels, from all over the world, ranging from peasant farmers to the working class to professionally and politically prominent individuals. One characteristic they all have in common is emotional stress associated with a traumatic event. In 1994, the late Harvard psychiatrist and Pulitzer Prize–winner John Mack, MD (who prevailed in an attempt by Harvard Medical School to censure him), contracted psychologists to administer a full battery of psychometric tests to four of his 76 cases. He cited the expense and time-consumption as reasons for limiting formal psychological testing on more of his patients. Three tested in the normal range with no psychopathology. The other had already been hospitalized for emotional disturbance, but cause and effect could not be sorted out.

In the absence of a primary psychiatric disorder, experimental psychologists sought out alternative explanations for alien abduction. They reasoned that they could find the answer by conducting experiments to discover personality disorders among alleged abductees. Several studies done throughout a 30-year period hypothesized abductees might be fantasy prone or unable to clearly delineate the boundaries between real and imaginary events, but the results are equivocal at best.

Psychologists S.C. Wilson and T.X. Barber coined the term "fantasy prone personality" in 1981. They conjectured that 4 percent of the adult population spends most of its time engaged in magical thinking, and that 65 percent of fantasy-prone individuals believe their fantasies are real. Fantasy-prone subjects reported childhood experiences in which they fantasized their dolls and stuffed animals were real or in which they pretended they were someone else. Many reported having imaginary playmates or believing in fairies or guardian angels. Although fantasy play is common among children, according to Wilson and Barber, fantasy-prone individuals carry their childhood fantasy experiences into adulthood.

Wilson and Barber reported that many fantasy-prone subjects believe they have psychic ability, the ability to heal others, and out-of-body experiences. As we have seen in Chapter 12, there is statistically significant scientific evidence that some psi phenomena are real and may not represent a personality disorder. It appears that Wilson and Barber were ignorant of the fact that the CIA has worked extensively with psychic spies, who are able to remote-view enemy installations. The idea that these unusual perceptual abilities represent fantasy-proneness is not in accordance with scientific findings.

Wilson and Barber stated that 96 percent of their highly hypnotizable subjects were more fantasy prone than the general population. However, recent findings indicate hypnotizability relates more to an individual's ability to concentrate than to having an especially vivid imagination. In fact, empirical studies have shown that many imaginative subjects are poor candidates for hypnosis, and many highly hypnotizable subjects are not fantasy prone. Therefore, when we learn that a hypnotic subject is highly suggestible, it does not indicate that he is malleable or compliant. It means only that he is a good hypnotic subject. This is an important distinction.

Additional research designed to test the hypothesis that UFO abductees are fantasy prone has produced largely negative results, with one exception. Bartholomew, Basterfield, and Howard, in a 1991 study of alleged UFO abductees, based upon biographical reports of childhood fantasies, not psychometric tests, found that 87 percent had one or more of the major symptoms of fantasy-prone personality, primarily psychic experiences—again, a contentious finding. However, a 1990 empirical study using psychometric tests rather than biographical data, by Ring and Rosing, found that suspected abductees were no more fantasy prone than the control group. Additionally, a 1991 study by Rodeghier et al on subjects who met clearly defined criteria for an abduction experience found no difference in the Inventory of Childhood Imaginings score for alleged abductees and the control group.

Nicholas Spanos et al concluded that their finding indicate that those who report UFO experiences, even missing time and telepathic communication with aliens, are no more fantasy prone than the general population. So if alleged abductees aren't fantasy prone, what are they?

Martin Kottmeyer, a Midwestern farmer and vocal skeptic, proposed boundary deficit disorder as a possible explanation for alien abduction. Although he is not a behavioral scientist, his article gained wide acceptance within debunking groups and came to the attention of experimental psychologist Nicholas Spanos. Kottmeyer hypothesized, based upon the 1984 Hartmann study of nightmare sufferers, that alien abductees, in all probability, exhibit boundary deficit symptoms, such as difficulty in differentiating between fantasy and reality, poor sense of self, poor social adaptation with frequent feelings of rejection, suicidal tendencies, feelings of powerlessness, and unusual alertness to sights, sounds, and sensations. In 1993, Spanos et al administered five psychometric scales to a control group and to close-encounter subjects. Test results revealed that the experimental subjects exhibited lower schizophrenia, higher self-esteem, higher well-being, lower perceptual aberration, lower perception of an unfriendly world, lower aggression, and no difference from the control group in social potency.[4] These finding are diametrically opposed to Kottmeyer's hypothesis. He was wrong on every hypothesis. Additional testing revealed no difference between the control and experimental groups in absorption, fantasy-proneness, and the tendency to engage in imaginings.

Experimental psychologists have more recently focused upon individuals who report nocturnal bedroom abductions. Hundreds claim to have awoken paralyzed with a group of small figures standing beside their beds. Some attempted to cry out but could not vocalize their fright. Some who awakened without paralysis cried out and attempted to escape, but to no avail. Soon a wave of paralysis overtakes them, and they are whooshed from their homes to waiting craft where they reportedly are subjected to involuntary biological experiments. Some are returned to their beds, but others end up locked outside of their homes. Some awaken on their roofs, in their vehicles, or even in someone else's home.

Sleep paralysis has been advanced as a hypothetical explanation for nocturnal bedroom abduction. It is experienced by about 30 percent of the population at least once. Sleep anomaly experiencers think they are awake but are utterly paralyzed, except for their eyes. They report accompanying feelings of fear or anxiety, pressure on their chests, and shadowy figures in their rooms. Sleep paralysis is a normal function of REM sleep, because it protects us from acting out our dreams in a physical sense. But occasionally we emerge from sleep while the paralysis continues for a few seconds.

Some experimental psychologists hypothesize that extraterrestrial entities are generated in hypnagogic (between waking and sleeping) and hypnopompic (between sleeping and waking) sleep states, a condition that affects about 5 percent of the population. Hypnagogic and hypnopompic hallucinations occur when factors such as stress, extreme fatigue, medications, and mental illness cause the part of the brain that distinguishes between conscious perceptions and internally generated perceptions to misfire. This results in internally generated visions, sounds, feelings, smells, or tastes. Experiencers often see colored geometric shapes or parts of objects. Others might observe the colored image of a person, monster, or animal. Hypnagogic hallucinations can be frightening and often cause a sudden jerk and arousal just before falling asleep. The hallucinations can last from seconds to minutes and are usually accompanied by a momentary period of sleep paralysis. Hypnagogic and hypnopompic hallucinations occur at a high rate of frequency among narcoleptics, who experience extreme fatigue or periods of dozing off during the day, but most sleep-anomaly sufferers report they are able to differentiate between their internally generated hallucinations and reality.

Elizabeth Loftus, PhD, offers an alternative hypothesis: false memory syndrome. She initiated false memory studies in response to a frenzy of childhood sexual abuse charges, some of which were caused by suggestion from authority figures and were not based in reality. False-memory syndrome is defined as an experience such that people remember events that never happened to them as if they are memories of real events.

To test false memory formation, social research scientists conduct false memory experiments by presenting semantically related word lists that include such words as "sour," "candy," "sugar," "good," "taste," "tooth," "nice," and "honey" to groups of experimental subjects. There is also a critical lure word, such as "sweet," which is not included in the lists. The words are generally presented orally on an audio recording at a rate of one word every three seconds. Next, the subjects are distracted with an assignment such as simple math problems or a short reading assignment followed by questions. Finally, the subjects are asked to write the words they recalled. They are then presented the list with the critical lure and asked to identify the orally presented words. Test subjects who recognized the critical lure, "sweet," are identified as having developed a false memory for the incorrect word. The experimenters assume that this simple word-identification mistake can translate into false memories for complex experiences.

But not all researchers agree. Research at Colgate University on individual differences in the formation of false memories found that high suggestibility is not related to the formation of false memories in the semantic word association test.

The false memory research most vigorously contested by researchers who support the alien abduction hypothesis was conducted by Susan Clancy, Richard J. McNally, et al (2002, 2004) at Harvard University. It focused upon the formation of false memories in people reporting abduction by aliens. The researchers recruited test subjects through two advertisements in local publications. The experimental subject ad stated that Harvard University researchers were "seeking people who may have been contacted or abducted by space aliens to participate in a memory study."[5] The control group ads simply stated Harvard University researchers were "seeking people to participate in a memory study."[6] The researchers restricted their inclusion of experimental subjects to individuals who met the criteria for sleep paralysis and hypnagogic hallucinations of extraterrestrial aliens—not individuals who met the criteria for alien abduction (conscious memory of a close encounter with a UFO and/or alien beings while outside one's home, witnesses to the event, missing time, forensic evidence, consistent hypnotic recall by multiple witnesses, passing a polygraph exam, and testing within the normal range on psychological screenings). In fact, all potential recruits who reported conscious, continuous memories of alien abduction were excluded

204 Science Was Wrong

from the experiment! All alleged abductees had been exposed to popular media pertaining to alien abduction. The test subjects were divided into three groups: recovered memory, repressed memory, and a control group. They completed four subjective-experiences scales designed to measure post-traumatic stress disorder, depressive symptoms, memory lapses, and hypnotic suggestibility, and four schizotypy and schizophrenia measures. Additionally, participants completed a semantic word-association test, similar to the one previously described, designed to measure memory acuity and false memory formation.

Clancy and her team hypothesized that the experimental group would recall a higher percentage of false targets on semantically related word lists than the control group, suggesting that they were prone to false recall and false recognition. This hypothesis fell short of statistical significance on false recall but was significant on false recognition.

Additional tests indicated that although the recovered and repressed memory groups experienced a slightly higher degree of depressive symptoms and anxiety than the control group, they were, for the most part, normal, although many exhibited higher levels of creativity, vivid memory formation, open-mindedness toward psi experiences, and the ability to become absorbed in music, a movie, or nature, which Clancy et al interpreted as fantasy-proneness. These are characteristics of the learning style of right-brained, non-linear thinkers.

A critical analysis of the test subjects' scores reveals that the control group performed below the norm on various measures, including the Perceptual Aberration and Magical Ideation Scales. They seemed to be a group of particularly linear thinkers. If this observation is correct, as a group they would be expected to perform better on an orally presented memory test than would non-linear thinkers. We know that these left-brained linear types are auditory thinkers who process information in a sequential, analytical order. It appears that the test subjects were of two opposite learning-style types, and this would necessarily skew the test results.

The only conclusion we can draw from all of the social research findings is that fantasy-prone persons with thin boundaries, and those who experience certain sleep anomalies (narcolepsy), might believe they have been abducted by aliens, when they have not. If they are hypnotized by authority figures that suggest UFO abduction or firmly believe they have been abducted, they might confabulate an abduction experience and believe it is true. However, the following point is critically important: responsible abduction researchers and therapists refuse to hypnotize individuals who fall into this category. The primary requirement for allowing a person to undergo hypnosis is substantial evidence that the experience was *not* merely a hallucination or fantasy. A full 30 percent of abductees recall the entire experience without hypnosis, and do not fit nicely into the sleep-anomaly or fantasy-prone personality group. Credible witnesses, multiple-witness radar, and visual sightings sometimes involving skilled military personnel, as well as physical trace evidence cases support the idea that *some* individuals have been abducted. Many report that their initial experience occurred when they were driving, fishing, camping, and the like, or in the company of friends or relatives. Subsequent nocturnal bedroom abductions often follow.

A 2007 study by Tamara Lagrandeur, Don C, Donderi, Stuart Appelle, and abduction researcher Budd Hopkins examined a group of alleged alien abductees who had reported seeing alien symbols during an abduction experience. All symbol information was kept confidential and there was no possibility for information contamination. A separate group of graduate students was hypnotized and asked to imagine and illustrate symbols inside an alien spacecraft. The abductee group produced remarkably similar symbols, which were distinctly different from the symbols produced by the graduate students. The research scientists concluded that the consistency of symbols seen inside alien craft could possibly mean that the reports are true.

For many years, abduction researchers have attempted to gather physical evidence of alien abduction for laboratory analysis by highly reputable research scientists. Budd Hopkins, David Jacobs, Professor Emeritus Leo Sprinkle (forced from his job at the U. of Wyoming because of his alien abduction research), John Carpenter, John Mack, James Harder, and many others have cited numerous cases of physical and circumstantial evidence.

Approximately 40 percent of UFO abduction reports begin with the subjects fully awake, walking in the woods or driving in a car. Sometimes the alleged abductee is physically missing and being searched for by others. Witnesses to abductions observe a UFO within hundreds of feet of the abductee's home, and sometimes the craft leaves physical trace evidence on the ground. Missing persons are often returned dressed in unfamiliar clothing or find themselves in a heap outside their homes, which are deadbolt-locked from the inside. Additionally, some abduction accounts involve multiple witnesses who report detailed information when hypnotized separately.

Additionally, analyses of some purported landing circles have discovered strong electromagnetic fields and evidence of exposure to microwave radiation. Soil samples taken from the circles have exhibited unusual characteristics in comparison to nearby control samples. Sometimes heat exposure melts minerals within the soil, or the soil will no longer support vegetation or hold water.

In an attempt to secure physical evidence of abduction, hypno-anesthesia therapist/abduction researcher Derrel Sims proposed the surgical removal of alien implants. He made his first public presentation on alleged alien implants that had been surgically removed in 1992, following several years of investigation. In 1995, he presented the idea of surgically removing suspected implants, under the auspices of Saber Enterprises, to podiatrist Dr. Roger Leir, and in 1996, they assembled a surgical team for the first of several publicized procedures and analyses at top-notch laboratories. Today, both continue their research independently and have written books about their anomalous findings. Early research revealed that the objects are covered with a strange, gray, dense, biological membrane. The surrounding tissue exhibits no signs of inflammatory response, and a large number of specialized nerve cells found within the tissue samples had no anatomical reason to be there. At least one suspected implant removed by Leir's team emitted radio frequencies in extremely low frequency and microwave bands. One implant emitted a radio frequency that would broadcast in a deep-space radio frequency. Analytical tests performed on one of Leir's specimens showed trace element patterns and isotopic ratios consistent with meteoric origin.[7] Derrel Sims discovered dermal and subdermal fluorescence under a black ultraviolet light in 1992, which he postulates is a biological agent left on abductees as a residue. The laboratory report indicated the fluorescence on one suspected implant was caused by the metallic interior leaching through the dense biological outer membrane.[8]

《《《 》》》

The late Betty and Barney Hill. Archival photo from the Hill family collection. Courtesy of Kathleen Marden.

The September 19–20, 1961, Betty and Barney Hill UFO abduction in New Hampshire's White Mountains is one of the most credible cases of alien abduction. The couple observed an unconventional craft that was visible for more than 30 minutes and was viewed through binoculars during three observational stops. At one stop, in North Lincoln, they estimated that the large, silent, hovering disk was less than 100 feet above their vehicle. Barney left his car and walked into a field where he observed eight to 11 figures standing behind a row of windows through which a "bluish-white fluorescent glow shone." He consciously recalled observing figures dressed in "shiny black uniforms" that moved "smoothly and efficiently," as if to carry out a plan. In a confidential statement to NICAP investigator Walter Webb, he spoke of one figure's "intense concentration" and his observation that the figures were "somehow not human."[9]

The late Carl Sagan misinformed the public in his March 7, 1993 *Parade* article "Are They Coming For Us?" with this statement: "Betty spotted a bright star-like UFO that seemed to follow them. Because Barney feared it might harm them, they left the main highway for narrow mountain roads"[10]—despite the fact that Stanton Friedman faxed a 12-page correction letter to him on January 29, 1993 (two months prior to publication), after receiving a review copy of his manuscript. All accurate reports clearly state that Barney fled to his vehicle and hastened down U.S. Highway Route 3. The Hills informed Pease Air Force

Base that they assumed the UFO had apparently departed shortly after Barney resumed his drive along U.S. Highway 3, the major north–south highway through northern and central New Hampshire—not down narrow mountain roads. The Air Intelligence report states, "While the object was above them after it had 'swooped down' they heard a series of short loud 'buzzes' which they described as sounding like someone had dropped a tuning fork. They report that they could feel these buzzing sounds in their auto. They continued on their trip and when they arrived in the vicinity of Ashland, N.H., about 30 miles from Lincoln, they again heard the 'buzzing' sound of the 'object'; however, they did not see it at this time."[11]

It is difficult to understand why a prominent astronomer such as Carl Sagan would willfully misrepresent the Hill UFO encounter to more than 400 million viewers, yet the dramatization on his *Cosmos* television series depicts a nervous Betty gazing upward through torrential downpours with windshield wipers swishing, while at the same time she turns the knob on the radio to eliminate static. The truth is that the radio was off and the weather was clear. The Mount Washington Observatory's archival record states, "It appears the 19th and 20th were both beautiful late summer days atop the Rockpile. The evening and nighttime conditions for the 19th were quite tranquil...visibility was 130 miles throughout the night."[12] Sagan owed Betty Hill an apology, and his millions of readers and viewers an explanation, for his gross misrepresentation. Instead of correcting his false statements, he promoted them to the scientific community, and they have been repeated by misinformed social scientists and debunkers who have failed to engage in unbiased research about the Hill abduction case.

An apparent two-hour period of missing time, which defied prosaic explanation after investigation, and haunting memories of his experience ultimately led Barney to seek out the services of a psychiatrist. Following his UFO experience, which left physical evidence that something inexplicable had occurred, Barney developed high blood pressure, bleeding ulcers, nervousness, and insomnia; 27 months later, he was referred to Dr. Benjamin Simon, a prominent psychiatrist celebrated for his successful treatment of World War II veterans suffering from post-traumatic stress disorder. He knew almost nothing about UFOs, but was skilled at deep trance hypnotherapy to resolve psychogenic issues related to traumatic amnesia. He tape-recorded 10 individual, rather astonishing sessions with Betty and Barney, which continued with weekly sessions for six months. To guarantee privacy, Dr. Simon played bombastic classical music in the waiting room. At the end of each session, he induced amnesia in both Betty and Barney to

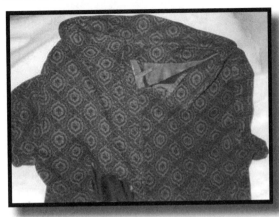

Evidence of damage to the zipper area of the dress Betty Hill was wearing on her September 19, 1961 trip from Montreal, Canada, to Portsmouth, N.H. Courtesy of Kathleen Marden.

ensure that no cross-contamination of information would occur, and to protect his patients' psychological well-being—he reasoned that if the traumatic event, for which they had amnesia, entered consciousness too early, it could produce added trauma.

Hypnotic regression brought forth nearly identical memories of a close encounter with non-human entities and a subsequent abduction in which the couple underwent a physical examination. (Read *Captured! The Betty and Barney Hill UFO Experience* by Stanton Friedman and Kathleen Marden for additional details.)

A scientific investigation of the Hill case reveals the following:

- There were two primary witnesses.
- There was conscious recall of CE-I (which signifies a craft within 500 feet of witness—the Hills reported it was approximately 100 feet away), a CE-II (beeping sounds and tingling), and a CE-III (Barney saw occupants on craft).
- There was conscious recall of missing time (the Hills arrived home two hours later than anticipated).
- There was physical evidence (Betty's badly torn dress coated with a pink powdery substance that to this day

has not been identified despite several chemical analyses, scraped shoe tops, magnetized shiny spots on the trunk of the car that caused the needle on a compass to whirl, and broken watches).

- There was consistent hypnotic recall of CE-IV. (Under hypnosis they "relived" a nearly identical experience. Both reported correlating, detailed positional and numerical information regarding their abduction experience that was not in a series of five dreams/nightmares experienced by Betty shortly after the close encounter experience. Because the Hills' amnesia remained intact during their hypnosis sessions, they could not have contaminated each other's recall.)[13]

- Both tested as normal in psychological screening (ACL), Barney as normal with strong intellectual talents, and Betty as normal but more independent, energetic, and poised than most.

- Both were deemed as not suffering from mental illness by psychiatrists.

- Both had excellent character references.

- Neither had a prior interest in UFOs.

- Betty passed a polygraph exam after Barney's death. Barney was not tested.

The Hill case is one of several credible UFO close encounter/abduction cases that leads scientists who research and investigate the subject to believe that at least some Earthlings have encountered non-human entities. Despite pleas to the scientific establishment to conduct objective studies of the evidence, cooperation has not been forthcoming. Additionally, scientific research articles that cast UFO reality in a positive light are often rejected by mainstream scientific journals.

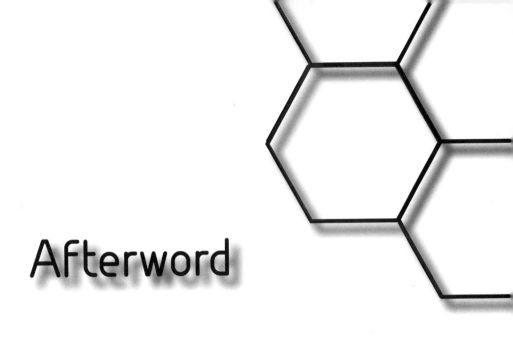

Afterword

It is clear that claims of impossibility have often been responsible for considerable damage to people and slowed technological progress. Areas often deemed impossible include a number of so-called paranormal phenomena such as near-death experiences, reincarnation, telepathy, remote viewing, and flying saucers. Claims have been made for the impossibility of government cover-ups, despite unassailable evidence that many secrets, even involving billions of dollars, have indeed been kept from the public. Stem-cell research and nanotechnology have been attacked. Nutritional claims often seem to be attacked because they might lead to fewer sales by pharmaceutical companies.

In the end, opposition to new findings only serves to hinder progress in fields of science. Surely, we need to avoid jumping to conclusions merely because well-educated people make false claims of impossibility.

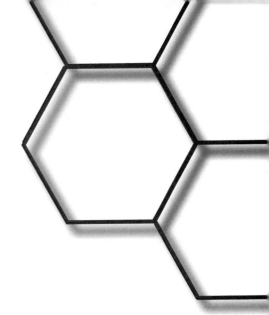

Notes

1

1. Lord Kelvin quoted from letter to Lord Baden Powell, December 8, 1896.

2. H.G. Wells, *Anticipations*, 208.

3. Simon Newcomb, "The Outlook," 2509.

4. Thomas Edison in *New York World* (Nov. 17, 1895) Cited in Christopher Cerf and Victor Navasky's *The Experts Speak* (New York: Pantheon Books, 1984), 236.

5. Wilbur Wright quoted in *The Book of Predictions*.

6. Worby Beaumont quoted in *Don't Quote Me*, 152. Cited in Cerf and Navsky, *The Experts Speak*.

7. George Melville in "The Engineer," 830–831. Cited in Cerf and Navsky, *The Experts Speak*, 238.

8. Octave Chanute, *Popular Science Monthly* (March 1904): 393.

9. Charles Stewart Rolls quoted in *Unpublished Collection of Unfortunate Predictions* (1908). Cited in Cerf and Navsky, *The Experts Speak*, 237.

10. Newton Baker quoted in *Billy Mitchell* (New York: E.P. Dunlop, 1942)

11. Josephus Daniels quoted in *Billy Mitchell.*

12. William H. Pickering quoted in *Aeronautics* (1908). Cited in Cerf and Navsky, *The Experts Speak*, 245.

13. *Scientific American* (July 16, 1910).

14. Marechal Foch quoted in *Coronet* (August 1914). Cited in Cerf and Navsky, *The Experts Speak*, 245.

15. John Wingate Weeks quoted in *Billy Mitchell*, 74.

16. Arlington B. Conway quoted in *American Mercury* (February 1932). Cited in Cerf and Navsky, *The Experts Speak*, 245.

17. John W. Thomason quoted in *American Mercury* (1937): 378. Cited in Cerf and Navsky, *The Experts Speak*, 245.

18. Einstein quoted in *Hiroshima Plus 20*, 76.

19. Lord Ernest Rutherford quoted in *Physics Today* (October 1970). Cited in R.L. Weber and E. Mendoza, *Random Walk in Science* (Philadelphia: Keyden & Son, 1973), 131.

20. Harry S. Truman, *Memoirs, Volume 1: Year of Decisions* (Garden City: Doubleday and Co., 1955), 11.

2

1. *New York Times.* January, 13 1920.

2. *New York Times.* July 17, 1969.

3

1. Fred Hoyle, *Nature of the Universe* (New York: Harper, 1960).

2. Valentine Bargmann and Lloyd Motz, "On the Recent Discoveries," 1350–52.

3. B.F. Burke and K.L. Franklin, *Journal of Geophysical Research*, 213–217.

4. V. Radhakrishnan and J.R. Roberts, *Physical Review Letters*.

5. Robert O. Becker and Gary Selden, *The Body Electric*.

6. "Flux Transfer Event Links Sun, Earth's Magnetic Field," FoxNews.com.

7. *FATE Magazine*. The Article can be read at *http://cura.free.fr/ xv/14starbb.html*. Quote is from the cover page. A more comprehensive 69-page overview of the controversy was compiled by Jim Lippard, entitled "Skeptics and the Mars Effect: A chronology of Events and Publications"; see *http://www.discord.org/~lippard/mars-effect-chron.rtf*.

8. Paul Kurtz is a debunker, not a skeptic.

9. Carl Sagan, "Letter," *The Humanist* 36 (1976): 2.

10. David M. Jacobs, *The Threat*.

4

1. Benjamin, Louise, "In Search of the Sarnoff 'Radio Music Box' Memo," 97–106.

2. Atyeo and Green, *Don't Quote Me*, 154.

3. *Literary Digest* (Nov. 6, 1926): 9.

4. Chris Morgan and David Langford, 20.

5. Gabe Essoe, *Book of Movie Lists*, 222.

6. Robert Conot, *Streak of Luck*, 245.

7. T. Craven quoted in "Top 50 Failed Technology Predictions of All Time," *http://data-katalog.com/index.php?newsid=50975*.

8. *Popular Mechanics*, (March 1949): 258. Cited in Cerf and Navsky, *The Experts Speak*, 208.

9. Cited in Cerf and Navsky, *The Experts Speak*, 209.

10. "The Von Neumann Architecture of Computer Systems"; see *http://virtuallearning.ning.com/forum/topics/the-von-neumann-archictecture*.

11. Jack B. Rochester and Donald T. Gantz, *Naked Computer*.

6

1. Ole Daniel Enersen, "Ignaz Philipp Semmerweis," *www.whonamedit.com/doctor.cfm/354.html.*

2. Ibid.

3. Ibid.

4. "Semmelweis: Defender of Motherhood," *http://dodd.cmcvellore.ac.in/hom/26%20-%20Semmelweis.html.*

5. Ibid.

6. Sherwin B. Nuland, *The Doctor's Plague*, 167–168.

7. Ibid., 160–161.

8. Ibid.

7

1. "Native Blood: The Myth of Thanksgiving," *Revolutionary Worker*, No. 833 (Nov. 24, 1996).

2. "Jeffrey Amherst and the Smallpox Blankets."

3. Ibid.

4. Ibid.

5. *The American Heritage Dictionary*, 1338.

6. Nicolau Barquet, MD and Pere Domingo, MD, "Smallpox."

7. G.C Kohn, editor. *Encyclopedia of Plague and Pestilence*, through *www.dshs.state.tx.us/preparedness/bt_public_history_smallpox.shtm.*

8. Andrew D. White, *A History of Warfare of Science with Theology*, 56.

9. *The American Heritage Dictionary*, 1334.

10. Jonathan B. Tucker. *Scourge*, 24.

11. Edward M. Crookshank, *History and Pathology of Vaccination*, quoted from Christopher Cerf and Victor Navasky, *The Experts Speak*, 33.

12. Charles Creighton, *Jenner and Vaccination*, quoted from Christopher Cerf, Victor Navasky, *The Experts Speak*, 33.

8

1. *Sanderson* is a pseudonym for a real family who wishes to remain anonymous. Certain details about their lives have been fictionalized to protect their identity.

2. Randy Shilts, *And the Bad Played On*, 115.

3. Ibid., 151.

4. Ibid., 57.

5. Andre Picard, *The Gift of Death*, 70.

6. Shilts, 242–243.

7. Shilts, 307.

8. Ibid.

9. John Crewdson, "In Gallo Case Truth Termed a Casualty," *Chicago Tribune* (January 1, 1995).

10. Ibid.

11. "Discoverers of HI Virus Win Share of Nobel Medicine Prize," Mail and Guardian Online, October 6, 2008.

12. Picard, 100.

9

1. Letter from Charles Davenport to V.L. Kellogg, October 30, 1912.

2. Francis Galton, *The American Journal of Sociology* X:1 (July 1904).

3. Statement made to British naturalist William Bateson in 1904 defending his *Index to Achievements of New Kinfolk*. Edwin Black, *War Against the Weak*, 26, 28.

4. John H. Noyes, "Essay on Scientific Propagation," Sections 2 and 15. Quoted from Black, *War Against the Weak*, 39.

5. Ibid. 44.

6. Harry H Laughlin, "Calculations."

7. Dr. F.W. Hatch, Superintendent of State Hospitals. Report of the State Commission in Lunacy for California. June 30, 1912. Cited by Harry Laughlin.

8. "Origins of Eugenics," *www.hsl.virginia.edu/historical/eugenics/2-origins.cfm.*

9. John Harvey Kellogg, "Eugenics and Immigration Needed."

10. Ibid., 433.

11. Ibid., 440. Citing the findings of Dr. Tredgold, the Registrar-General of England.

12. Ibid. 440–441.

13. Charles B. Davenport, PhD, "The Importance to the State of Eugenic Investigation."

14. Ibid., 454.

15. Harry H. Laughlin, 478.

16. Ibid., 450.

17. Ibid., 480.

18. Ibid., 2.

19. Buck V Bell 274 U.S. 200 (1927).

20. Foster Kennedy "The Problem of Social Control," 13–16.

10

1. Hightower, Jane M., MD. *Diagnosis Mercury*, 7.

2. "Minamata Disease," *http://www.absoluteastronomy.com/topics/Minamata_disease.*

3. Hajime Hosokawa, director of Chisso Hospital. Patient records.

4. "Minamata Disaster" *TED Case Studies #246.*

5. "Minamata Disease."

6. Ibid.

7. Masazumi Harada, *Minamata Disease*, 52.

8. Hightower, 141.

9. Farham Bakir, et al. "Clinical and Epidemiological Aspects of Methylmercury Poisoning," 2.

10. Ibid., 6.

11. Sanfeliu, et al. "Neurotoxicity of Organomercurial Compounds," 3.

12. "A Case Study: The Toxic Legacy of the California Gold Rush." International Indian Treaty Council.

13. "Endangered Great Lakes," *http://www.time.com/time/magazine/article/0,9171,943821,00.html.*

14. John Schertow, "Mad as a Hatter," 3.

15. John Schertow, "Great Lakes: Chemical Soup."

16. John Scertow, "Court Rules EPA Violated the Law."

17. See FDA mercury levels in commercial fish online at *www.cfsan.fda.gov/~frf/sea- mehg.html.*

18. Gerald Rellick, "George Bush on Mercury Pollution."

11

1. Ingrid Dengel, Dominick Aeby, and John Grace, "Relationship Between Galactic Cosmic Radiation and Tree Rings," 184: 545–551.

2. Orinn G. Hatch quoted in "UN Climate Scientists Speak Out on Global Warming."

3. Frederick Seitz quoted in "A Major Deception in Global Warming," *The Wall Street Journal* (June 12, 1996). Cited in Craig Idson and S. Fred Singer's "Climate Change Reconsidered: 2009 Report on the Nongovernmental Panel on Climate Change." (NIPCC)

4. Ibid. 8, 751.

5. Habibulto Abdussamatov. "The Sun Defines the Climate."

12

1. Brian Josephson, "Scientists' Unethical Use of Media for Propaganda Purposes." University of Cambridge. *www.tcm.phy.cam.ac.uk/~bdj10/propaganda/index.html.*

2. Andrew A. Skolnick, "Natasha Demkina."

3. Brian Josephson, "Scientists' Unethical Use of Media for Propaganda Purposes."

4. Phil Baty, "Scientists Fail to See Eye to Eye Over Girl's X-ray Vision," 3.

5. Ibid., 4–5.

6. Victor Zammit. "Natasha Demkina Can Sue the 'Ambush Experimenters.'"

7. Dean Radin. *The Conscious Universe*, 65.

8. Charles Honorton and Ray Hyman, "The Ganzfeld PSI Experiment," 351.

9. C. Holden. "Academy Helps Army Be All That It Can Be," *Science* 238 no. 4833 (Dec. 11, 1987): 1502. Cited in Radin, *The Conscious Universe*, 215.

10. Jessica Utts, "Replication and Meta-Analysis in Parapsychology," 363–404.

11. Ray Hyman. Comment. *Statistical Science* 6 (1991): 392.

12. Radin, *The Conscious Universe*, 88.

13

1. Memo from Robert Low to James Archer, the dean of the University of Colorado, dated 8/9/1966.

2. Stanton T. Friedman, *Top Secret/Majic, 2nd Edition*.

3. New Webster's Dictionary.

4. For additional information read *Shoot Them Down! The Flying Saucer Wars of 1952* by Frank C. Feschino, Jr. (Lulu Publications, 2007).

5. Memo from Brigadier General Carroll H. Bolender, *Unidentified Flying Objects* October 20, 1969.

14

1. Budd Hopkins, *Art, Life and UFOs: A Memoir*, 361.

2. National Academy of Sciences, "Review of the University of Colorado Report on Unidentified Flying Objects."

3. Memo from Robert Low to James Archer, the Dean of the University of Colorado, dated 8/9/1966.

4. Ibid., 40.

5. Susan A. Clancy, et al. "Memory Distortion in People Reporting Abduction by Aliens," 456.

6. Ibid.

7. Dr. Roger Leir, *Casebook: Alien Implants*, and lecture by Roger Leir at the 2009 International UFO Congress and Film Festival.

8. Derrel Sims, *Alien Hunter*. Also, personal correspondence between Marden and Sims.

9. Walter Webb, "A Dramatic UFO Encounter in the White Mountains, NH." Confidential NICAP Report, October 26, 1961.

10. Carl Sagan. "Are They Coming for Us?" *Parade,* 5.

11. See Chapter 3, "The Project Blue Book Report," and the Appendix of *Captured! The Betty and Barney Hill UFO Experience* by Friedman and Marden.

12. Correspondence with Tim Markle, chief meteorologist at the Mount Washington Observatory.

13. For the full magazine article see *The MUFON UFO Journal,* April 2009 issue.

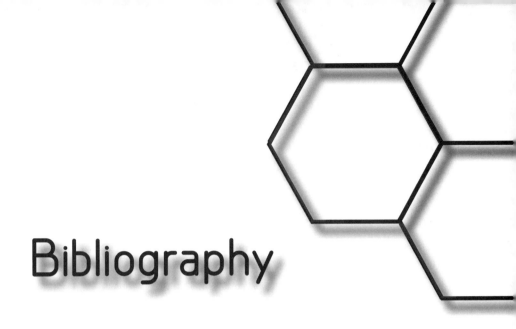

Bibliography

1

Atyeo, Don, and Jonathon Green. *Don't Quote Me: What People Said— and Then Wished They Hadn't*. London: Chancellor Press, 1994.

Conway, Arlington B. "Death From the Sky." *American Mercury*. February 1932.

Edison, Thomas. *New York World*, Nov. 17, 1895.

Finney, John. *Hiroshima Plus 20*. New York: Delacorte Press, 1965.

Foch, Marechal. *Coronet*. August 1914.

Gauvreau, Emile, and Lester Cohen. *Billy Mitchell*. New York: E.P. Dunlop, 1942.

Lord Kelvin. Letter to Lord Baden Powell. December 8, 1896.

McCullocgh, David. *Truman*. New York: Simon and Shuster, 1992.

Melville, George. "The Engineer and the Problem of Aerial Navigaton." *North American Review*. December 1901.

Newcomb, Simon. "The Outlook for the Flying Machine." *The Independent*, October 22, 1903.

Pickering William H. *Aeronautics*. 1908.

Rutherford, Lord Ernest. *Physics Today*. October 1970.

Scientific American July 1910.

Sheaffer, Louis. "Unpublished Collection of Unfortunate Predictions."

Wallechinsky, David, and Amy Wallace. *The Book of Predictions*. New York: William Morrow and Co, 1981.

Weber, R.L. and E. Mendoza. *Random Walk in Science*. (Philadelphia: Keyden & Son, 1973).

Wells, H.G. *Anticipations of the Reaction of Mechanical and Scientific Progress Upon Human Life and Thought*. New York: Harper and Brothers, 1902.

2

Cameron, Alistair G.W. *Interstellar Communication: The Search for Extraterrestrial Life*. New York and Amsterdam: W.A. Benjamin, Inc. 1963.

Campbell, John W., PhD. "Rocket Flight to the Moon." *Philosophical Magazine 7* (31), no. 204 (January 1941): 24–34.

Dyson, Freeman. "Gravitational Machines" *Interstellar Communication: The Search for Extraterrestrial Life*. New York and Amsterdam: W.A. Benjamin. Inc. 1963.

Goddard, Robert Hutchings, PhD. "A Method of Reaching Extreme Altitudes." *Smithsonian Miscellaneous Collections* vol. 71, no. 2., 1919.

Krauss, Lawrence Maxwell. *Beyond Star Trek: Physics From Alien Invasions to the End of Time*. New York, NY: Basic Books, 1997.

———. "Odds Are Stacked When Science Tries to Debate Pseudoscience." *The New York Times,* April 30, 2002.

———. *The Physics of Star Trek*. New York: Basic Books, 1995.

Luce, J. S. and John Hilton. "Controlled Fusion Propulsion." *Proceedings of 3rd Symposium on Advanced Propulsion Concepts* Vol. 1. New York: Gordon and Breach Science Publishers, 1963.

Purcell, Edward. "Radioastronomy and Communication Through Space." Brookhaven Lecture Series no. 1. Upton, New York: Brookhaven National Laboratory, 1962.

Time Magazine LXVII no. 3 (January 16, 1956).

"Topics of the Times: Robert Goddard" *New York Times* January 13, 1920.

Tsiolkovsky, Konstantin. "Exploration of Cosmic Space by Means of Reaction Devices." 1903.

von Hoerner, Sebastian. "The General Limits of Space Travel. *Interstellar Communication: The Search for Extraterrestrial Life.* New York and Amsterdam: W.A. Benjamin, Inc., 1963.

3

Bargmann, Valentine, and Lloyd Motz. "On the Recent Discoveries Concering Jupiter and Venus." *Science* 138 (December 21, 1962): 1350–52.

Becker, Robert O., and Gary Selden. *The Body Electric: Electromagnetism and the Foundation of Life.* New York: Morrow, 1998.

Bryner, Jeanne. "Flux Transfer event links Sun, Earth's Magnetic Field." FoxNews.com, Nov. 5, 2008. *www.foxnews.com/story/0,2933,447214,00.html.* Accessed 10/23/2009.

Burke, B.F. and K.L. Franklin. "Observations of a Variable Radio Source Associated with the Planet Jupiter" *Journal of Geophysical Research* 60 (1955): 213–17.

de Grazia, Alfred. *The Velikovsky Affair.* London: Sidgwick and Jackson, 1966.

FATE Magazine 34 (October 1981): 67–98. *http://cura.free.fr/xv/14starbb.html.* Accessed 9/16/2009.

Goldsmith, Donald, ed. *Scientists Confront Velikovsky.* New York: W.W. Norton Co., 1977.

Hoyle, Fred. *Nature of the Universe.* London: Blackwell, 1950.

Jacobs, David M. *The Threat: Revealing the Secret Alien Agenda.* New York: Simon and Shuster, 1998.

Lippard, Jim. "Skeptics and the Mars Effect: A Chronology of Events and Publications." *www.discord.org/~lippard/mars-effect-chron.rtf.* Accessed 8/28/2009.

Radhakrishnan, V., and J.R. Roberts. "Polarization and Angular Extent of the 960-Mc/sec Radiation from Jupiter." *Physical Review Letters* 4 (10): 493–94

Velikovsky, Immanuel. *Ages in Chaos.* 1952. Garden City, New York: Doubleday & Co., 1952.

———. *Earth in Upheaval.* Garden City, New York: Doubleday & Co., 1955.

———. *Worlds in Collision.* Garden City, New York: Doubleday & Co., 1950.

4

Benjamin, Louise. "In Search of the Sarnoff Radio Music Box Memo." *Journal of Radio and Audio Media* 9:1 (May 2002): 97–106.

Cerf, Christopher, and Victor S. Navasky. *The Experts Speak: The Definitive Compendium of Authoritative Misinformation.* New York: Villard Books, 1998.

Conot, Robert. *Streak of Luck: The Life and Legends of Thomas Alva Edison.* New York: Seaview Books, 1979.

Essoe, Gabe. *Book of Movie Lists.* Westport, Conn.: Arlington House, 1981.

Morgan, Chris, and David Langford. *Facts and Fallacies: A Book of Definitive Mistakes and Misguided Predictions.* Exeter, England: Webb and Bower, 1981.

Popular Mechanics (March 1949).

Rochester, Jack B., and Donald T. Gantz. *Naked Computer.* London: Arlington Books Publishers Ltd., 1981.

Von Neumann "The Von Neumann Architecture of Computer Systems." *http://virtuallearning.ning.com/forum/topics/the-von-neumann-archictecture.* Accessed 6/14/2009.

5

Book review of *The Rebirth of Cold Fusion: Real Science, Real Hope, Real Energy* in *Journal of Scientific Exploration* 19, no. 2 (Summer 2005): 288–93.

Krivit, Steven B., and Nadine Winocur. *The Rebirth of Cold Fusion: Real Science, Real Hope, Real Energy.* Pasadena, Calif.: Pacific Oaks Press, 2004.

Mallove, Eugene. "Critical Review Dissects Voodoo Science." *Infinite Energy Magazine* 5, no. 30 (March/April 2000): 44.

Park, Robert L., PhD. *Voodoo Science: The Road From Foolishness to Fraud.* New York: Oxford University Press, 2000.

6

Colyer, Christa. "Childbed Fever: A Nineteenth-Century Mystery." University of Ontario Institute of Technology. *www.sciencecases.org/childbed_fever/childbed_fever.asp.* Accessed 10/2008.

Cwikel, Julie, Ph.D. "Lessons from Semmelweis: A Social Epidemiologic Update on Safe Motherhood." *Social Medicine.* 3:1 (2008): 19–35.

De Costa, C.M. "The Contagiousness of Childbed Fever: A Short History of Puerperal Sepsis and Its Treatment." *The Medical Journal of Australia* 177, no. 11–12 (2002): 2–16.

Enerson, Daniel Ole. "Ignaz Philipp Semmelweis." *www.whonamedit.com/doctor.cfm/354.html.* Accessed 10/13/2008.

Gilder, S.S.B. "Carl Von Rokitansky." *Canad. M.A.J.* 71 (July 1954). *www.pubmedcentral.nih.gov/picrender.fcgi?artid=1825066&blobtype=pdf.* Accessed 10/2008.

Hauzman, Erik E., "Semmelweis and His German Contemporaries." Budapest, Hungary: Semmelweis University Faculty of Medicine, Baross u. 27.

Jay, Venita, MD, FRCPC. "Ignaz Semmelweis and the Conquest of Puerperal Sepsis." College of American Pathologists. *www.archivesofpathology.org/doi/full/10.1043/0003-9985(1999)123%3C0561:ISATCO%3E2.0.CO;2.* Accessed 10/10/2008.

Nuland, Sherwin B. *The Doctors' Plague: Germs, Childbed Fever, and the Strange Story of Ignac Semmelweis.* New York: W.W. Norton and Sons, 2003.

"Revolutions of 1848–49." *http://histclo.com/essay/war/rev-1848.html.*3 Accessed 10/20098.

"Semmelweis: Defender of Motherhood." *http://dodd.cmcvellore.ac.in/hom/26%20-%20Semmelweis.html.* Accessed 10/16/2008.

Tan, S.Y., MD, JD, and J. Brown. "Medicine in Stamps: Ignac Philipp Semmelweis (1818–1865): Handwashing Saves Lives." *Singapore Medical Journal,* 47 no. 1 (2006): 6.

7

Aronson, Stanley M. and Lucile Newman. "God Have Mercy on This House: Being a Brief Chronicle of Smallpox in Colonial New England." *www.brown.edu/Administrations/News_Bureau/2002-03/02-017t.html.* Accessed 11/20/2008.

Barquet, Nicolau, MD, and Pere Domingo, MD, "Smallpox: The Triumph Over the Most Terrible of the Ministers of Death." *Annals of Internal Medicine* 127:6 (October 15, 1997): 635–42. *www.annals.org/cgi/content/full/127/7_Part_1/635.* Accessed 11/18/2008.

"Conquistodors-Amazonia." *www.pbs.org/opb/conquistodors/peru/adventure1/b2htm.* Accessed 9/1/2009.

Creighton, Charles. *Jenner and Vaccination.* London: Sonnenschein, 1889.

Crookshank, Edward M. *History and Pathology of Vaccination.* London: Lewis, 1889.

Faria, Miguel A., Jr., MD. "Jenner, Pasteur, and the Dawn of Scientific Medicine." *Science* (June 2001). *www.capmag.com/objective-science/ articles/mf_vaccines1a.htm*. Accessed 11/26/2008.

Flight, Colette. "Smallpox: Eradicating the Scourge." *www.bbc.co.uk/ history/britixh/empire_seapower/smallpox_print.html*. Accessed 11/28/2008.

"Germ Theory of Disease." *Nation Master Encyclopedia Online. www .nationmaster.com/encyclopedia/Germ-theory-of-disease*. Accessed 11/26/2008.

Henderson, Donald A., and Bernard Moss. "Smallpox and Vaccinia." *www.ncbi.nlm.nih.gov/bookshelf/br.fcgi?book=vacc&part=A3*. Accessed 9/23/2009.

"History of Smallpox—Smallpox Through the Ages." Texas Department of State Health Services. *www.dshs.state.tx.us/preparedness/bt_ public_history_smallpox.shtm*. Accessed 9/25/2009.

"Jeffrey Amherst and the Smallpox Blankets" *www.nativeweb.org/pages/legal/amherst/lord_jeff.html*. Accessed 9/1/2009.

Jenner Museum Online, The. *www.jennermuseum.com*. Accessed 9/23/2009.

Kohn, G.C., ed. *Encyclopedia of Plague and Pestilence*. New York: Facts on File, Inc., 1995. *www.dshs.state.tx.us/preparedness/bt_public_ history_smallpox.shtm*. Accessed 9/2009.

"Native Blood: The Myth of Thanksgiving." *Revolutionary Worker* no. 833 (November 24, 1996).

Reidel, Stefan, MD, PhD. "Edward Jenner and the History of Smallpox and Vaccination." *Baylor University Medical Center Proceedings* 18 no. 1 (January 2005): 21–25 *www.pubmedcentral.nih.gov/ articlerender.fcgi?artid=1200696*. Accessed 11/28/2008.

Reimer, Terry. "Smallpox and Vaccination in the Civil War." *www.civilwrrmed.org/Research/Articles.aspx*. Accessed 9/25/2009.

The American Heritage Dictionary. Boston: Houghton Mifflin Co., 1982.

"Smallpox and History." *www.cdc.gov/agent/smallpox/training/overview/ pdf.eradicationhistory.pdf*. Accessed 8/24/2009.

"Smallpox Through History." *http://encarta.msn.com/media_701508643/ smallpox_through_history.html*. Accessed 11/28/2008.

Sy, Tan, MD, JD, and l. Rogers, MD. "Louis Pasteur: The Germ Theorist. *Singapore Medical Journal* 48(1) (2007): 4.

"Theological Opposition to Inoculation, Vaccination, and the Use of Anaesthetics." *http://abob.uga.edu/bobk/whitem10.html*. Accessed 11/26/2008.

Tschanz, David, W., PhD. "The War Against Smallpox." *www.stratefypage.com/articales/smallpox/3.asp*. Accessed 9/9/2009.

Tucker, Jonathan B. *Scourge: The Once and Future Threat of Smallpox*. New York: Grove Press, 2001.

White, Andrew D. "Zabdiel Boylston." *A History of Warfare of Science with Theology*. New York: Dover Publications, Inc., 1960. *www.todayinsci.com/B/Boylston_Zabdiel.htm*. Accessed 11/26/2008.

8

Altman, Lawrence K. "French Sue U.S. Over AIDS Virus Discovery." *New York Times* December 14, 1985. *www.nytimes.com/1985/12/14/ world/french-sue-us-over-aids-virus-discovery.html*. Accessed 11/13/2008.

———. "U.S. Delays Licensing Blood Test to Detect AIDS." *New York Times* February 15, 1985. *www.nytimes.com/1985/02/15/us/ us-delays-licensing-blood-test-to-detect-aids.html*. Accessed 11/13/2008.

"Aminocaproic Acid." *Medline Plus Health Topics. http://www.ncbi. nlm.nih.gov/bookshelf/br.fcgi?book=meds&part=a608024*. Accessed 11/11/2008.

"Archive Report: Science subverted in AIDS dispute." *Chicago Tribune* January 1, 2006. *www.ChicagoTribune.com/news/nationworld/ chi-100608-hiv-discovery-nobel-prize*. Accessed 11/13/2008.

"Bayer Exposed HIV Contaminated Vaccine." MSNBC report *www.youtube.com/warch?v=Wf2u7j9FeVl*. Accessed 3/7/2009.

Blake, L.V. "Hemophilia and HIV—as hidden majority." International Conference on AIDS. New England Hemophilia Association. July 19–24, 1992. *http://gateway.nlm.nih.gov/MeetingAbstracts/ ma?f=102200766.html*. Accessed 10/21/2008.

Chase, Marilyn. "French Scientists Sue U.S. on AIDS Research, Royalties." *Wall Street Journal* December 16, 1985. *www.aidsinfoobbs.org/articles/wallstj/85/69.txt*. Accessed 11/15/2008.

Crewdson, John. "In Gallo Case, Truth Termed a Casualty." *Chicago Tribune* January 1, 1995. *www.virusmyth/aids/hiv/jcgallocase.htm*. Accessed 11/15/2008.

"Current Trends Changing Patterns of Acquired Immunodeficiency Syndrome in Hemophilia Patients—United States." *MMWR Weekly*. CDC 34/17 (May 3, 1985): 241–43. *www.cdc.gov/mmwr/preview/mmwrhtml/00000535.htm*. Accessed 10/21/2008.

"Discoverers of HI Virus Win Share of Nobel Medicine Prize." Mail and Guardian Online October 6, 2008. *http://www.mg.co.za/article/2008-10-06-discoverers-of-hi-virus-win-share-nobel-medicine-prize*. Accessed 11/13/2008.

Evatt, B.L. "The Tragic History of AIDS in the Hemophilia Population, 1982–1984." *Journal of Thrombosis and Haemostasis* 4/11 (September 14, 2006): 2295–2301. *www3.interscience.wiley.com/cgi-bin/fulltest/118577261/main.html,ftx_abs*. Accessed 10/21/2008.

Flieger, Ken. "Outlook Brighter for Youngsters With Hemophilia." *FDA Consumer Magazine* July/August 1993. *www.fda.gov/bbs/topics/CONSUMER/con00240.HTML*. Accessed 10/21/2008.

Ghosh, K. S Shetty, F. Jijina, D. Mohanty. "Role of epsilon amino caproic acid in the management of haemophilic patients and inhibitors." *Haemophilia* (2004). *www.ncbi.nlm.nih.gov/pubmed.14962221*. Accessed 11/11/2008.

"Global AIDS Timeline." *Kaiser Family Foundation* December 2007. *www.kff.org/hivaids/timeline/hivtimeline.cfm*. Accessed 10/21/2008.

Healy, Kieran. "The Emergence of HIV in the U.S. Blood Supply: Organizations, Obligations and the Management of Uncertainty." *Theory and Society* 28:4 (August 1999). *www.jstor.org/pss/3108561*. Accessed 10/21/2008.

"Hemophilia Litigation Media Center Press Articles." *www.Hemophilia-Litigation.com/media.htm*. Accessed 3/7/2009.

Hoey, John, MD. "Human Rights, Ethics, and the Krever Inquiry." *Canadian Medical Association* 157:9 (November 1, 1997).

Knox, Richard, and Steve Inskeep. "Nobel Prize in Medicine for Major Virus Discoveries." *NPR* (October 6, 2008). *www.npr.org/templates/story/story.php?storyId=95420499&ft=1&f=1004.* Accessed 11/13/2008

"Lewis K. Diamond 1902–1999." HemOnc Today January, 1 2008. *www.hemonctoday.com/articlePrint.aspx?type=print&rID=26031.* Accessed 11/11/2008.

McCarthy, Joyce. "New Test has Screened-Out AIDS Contaminants from Nation's Blood Supply, Expert Conferees Say." *The NIH Record* August 13, 1985.

Navarro, Mireya. "Hemophiliacs Demand Answers as AIDS Toll Rises." *New York Times* May 10, 1993. *http://query.nytimes.com/gst/fullpage.html?res=9F0CE7D6123DF933A25756C0A96595.* Accessed 10/21/1008.

Pasteur Institute. "Nobel Prize in Medicine 2008 Awarded to Professors Francoise Narre-Sinoussi and Luc Montagnier." *www.pasteur.fr/ip/easysite/go/03b-00002i-01f/press/press-releases/2008/nobel-prize-in-medicine-2008-awarded-to-professors-francoise-barre-sinoussi-and-luc-montagnier.* Accessed 12/17/2008.

Picard, Andre. *The Gift of Death.* Toronto: Harper Perennial, 1995.

"Risk Communication to Physicians and Patients." *HIV and the Blood Supply.* Institute of Medicine, National Academies Press (1995). *www.nap.edu/openbook.php?record_id=4989&page=169.* Accessed 10/21/2008.

"Robert Gallo." International Biopharmaceutical Associates. *www.ibpassociation.org/encyclopedia/medicine/Robert_Gallo.php.* Accessed 11/13/2008.

Roberts, Shauna S. "Blood Safety in the Age of AIDS." *Breakthroughs in Bioscience. http://opa.faseb.org/pdf/BloodSafety.pdf.* Accessed 10/21/2008.

Rock, Andrea. "America's Dangerous Blood Supply." *CNN Money.* May 1, 1994. *http://money.cnn.com/magazines/moneymag/moneymag_archive/1994/05/01/88826/index.htm.* Accessed 10/21/2008.

Schaefer, Judy, MA RNC. "Ethical Dilemmas in the Pediatric Hemophilia Community." *http://findarticles.com/p/articles/mi_m0FSZ/is_5_25/ai_n18609140/.* Accessed 10/21/2008.

Shilts, Randy. *And the Band Played On.* New York: St. Martin's Press, 1987.

Statement of Michael A. Stoto, PhD, for Congress. *The National Academies of Science* October 30, 1997. *www.nationalacademies .org/ocga/testimony/Aid%20HIV.asp.* Accessed 10/21/2008.

"Tainted Blood." *CBS News* February 27, 1998. *www.cbc.ca/news/background/taintedblood/.* Accessed 10/21/2008.

"Timeline of AIDS." *http://en.wikipedia.org.wiki/Timeline_of_AIDS.* Accessed 10/21/2008.

"Tranexamic Acid." *http://en.wikipedia.org.wiki/Tranexamic_acid.* Accessed 10/21/2008.

"Update: AIDS—United States, 2002." *MMWR Weekly, CDC* 51(27) (July 12, 2002): 592–95. *www.cdc.gov/mmwr/preview/mmwrhtml/ mm5127a2.htm.* Accessed 11/16/2008.

"What is Hemophilia?" *www.aolhealth.com/conditionas/hemophilia?fiv=1.* Accessed 10/20/2008.

9

Bateson, William. Statement made to British naturalist William Bateson in 1904 defending his *Index to Achievements of New Kinfolk.*

Black, Edwin. "Eugenics and the Nazis—The California Connection." SFGate.com. November 9, 2003.

———. *War Against the Weak.* New York: Four Walls Eight Windows, 2003.

———. "We Must Keep Eugenics Away From Genetics." *Newsday.com.* 10/15/2003. *www.waragainsttheweak.com/offSiteArchive/www. newsday.com/index.htm.* Accessed 11/29/2008.

Buck V Bell 274 U.S. 200 (1927). "Carrie Buck: Virginia's Test Case." *www.hsl.virginia.edu/historical/eugenics/3-buckvbell.cfm.* Accessed 5/11/2009.

Davenport, Charles. Letter to V.L. Kellogg, October 30, 1912.

Davenport, Charles B. PhD. "The Importance to the State of Eugenic Investigation." Proceedings of the First National Conference on Race Betterment. January 8–12, 1914. Battle Creek, Mich.

Galton, Francis. "Eugenics: Its Definition, Scope and Aims." *The American Journal of Sociology* X:1 (July 1904). *http://galton.org/essays/1900-1911/galton-1904-am-journ-coc-eugenics-scope-aims.htm*. Accessed 11/29/2008.

Gould, Stephen Jay. "Carrie Buck's Daughter: A popular, quasi-scientific idea can be a powerful tool for injustice." American Museum of Natural History (2002). *http://findarticles.com/p/articles/mi_m1134/is_6_111/ai_87854861*. Accessed 3/11/2009.

Hatch, F.W. Dr., Superintendent of State Hospitals. Report of the State Commission on Lunacy for California. June 30, 1912.

Holt, Marilyn Irvin. "Children's Health and the Campaign for Better Babies." *Kansas History: A Journal of the Central Plains* 28 (Autumn 2005): 174–87.

Jay, Joseph. "The 1942 Euthanasia Debate in the American Journal of Psychiatry." *http://sagepub.com/cgi/content/abstract/16/2/171*. Accessed 1/22/2009.

Johnson, Roswell Hill, and Paul Popenoe. "Applied Eugenics." *www.gutenberg.org/catalog/world/readfile?fk_files=W62273&pafeno=2*. Accessed 11/29/2008.

Kellogg, John Harvey. "Eugenics and Immigration Needed—A New Human Race." Proceeding of the First National Conference on Race Betterment. January 8–12, 1914. Battle Creek, Mich.

Kennedy, Foster. "The Problem of Social Control of the Congenital Defective: Education, Sterilization, Euthanasia. *American Journal of Psychiatry* 99 (July, 1942). *http://kpope.com/eugenics.php*. Accessed 1/22/2009.

Laughlin, Harry H. "Calculations on the working out of a Proposed Program of Sterilization." Proceedings of the First National Conference on Race Betterment. January 8–12, 1914. Battle Creek, Mich.

Lombardo, Paul. "Facing Carrie Buck." Hastings Center Report. March/April 2003.

Lovett, Laura L. "Fitter Families for Future Firesides: Popular Eugenics and the Construction of a Rural Family Ideal in the United States." U.M. Department of History. Amherst, MA. *www.cisd.yale.edu/agrarianstudies/papers/PopularEugenics.pdf*. Accessed 5/10/2009.

"Mendel, Gregor Johann." *Columbia Encyclopedia Online. www. referencecenter.com/ref/reference/Mendel-G/GregorJohann Mendel?invocation.* Accessed 5/21/2009.

Miller, Herbert Adolphus, PhD. "The Psychological Limit of Eugenics." Proceedings from the First National Conference on Race Betterment. January 8-12, 1914. Battle Creek, Mich.

Noyes, John H.. "Essay on Scientific Propagation." Oneida, N.Y.: Oneida Community, 1872.

"Origins of Eugenics." *www.hsl.virginia.edu/historical/eugenics/ 2-origins.cfm.* Accessed 5/19/2009.

Strode, A. "Letter to Harry Laughlin." Sept. 20, 1924. *www.eugenicsarchive.org/eugenics/image_header.pl?id=1336.* Accessed 5/21/2009.

10

"A Case Study: The Toxic Legacy of the California Gold Rush." International Indian Treaty Council. *www.treatycouncil.org/ IITCRTFAdvocacyTrainingpp-040-308.* Accessed 12/03/2008.

Aubrey, Allison. "The News Fuels Confusion about Women, Fish." *NPR* 1/11/2009. *www.npr.org/templates/story/story.php?storyId=15005507.* Accessed 1/11/2009.

Bakir, Farham, et. al. "Clinical and Epidemiological Aspects of Methylmercury Poisoning." *Postgraduate Medical Journal* (Jan 1980): 56.

"Bush Favors High Levels of Mercury." *Bush Watch* 3/31/05. *www.skeptically.org/bw/is3.html.* Accessed 1/8/2009.

Ceck, Paul, M.D., and Larry Wilson, M.D. "Mercury Toxicity." *www.arltma.com/MercuryToxDoc.htm.* Accessed 1/11/2009.

D'Itri, Frank M. "Mercury Contamination—What we have Learned Since Minamata." Institute of Water Research and Department of Fisheries and Wildlife. Michigan State University, East Lansing, Mich. *Environmental Monitoring and Assessment* 19 (1991): 165–82.

"Endangered Great Lakes." *http://time.com/time/print- out/0,8816,943821,00.html.* Accessed 1/12/2009.

"EPA Plan to Evade Required Cuts in Mercury Pollution Challenged in Court by States." Tribes, Health and Environmental Groups. *Environmental Defense* (2007). *www.bio-medicine.org/medicine-news-1/EPA-Plan-to-Evade-Required-Cuts-in-Mercury-Pollution-Challenged-in-Court-by-States--Tribes--Health-and-Environmental-Groups-7768-1/.* Accessed 1/7/2009.

Eto, Komyo, et al. "Reappraisal of the Historic 1959 Cat Experiment in Minamata by the Chisso Factory." *Tohoku Journal of Experimental Medicine* 194 (2001): 197–203.

"FDA's Midnight Mischief Heightens Mercury Risk to Pregnant Women, Infants." Environmental Working Group. 12/12/2008. *www.ewg.org/node/27431/print.* Accessed 1/7/2009.

Hosokawa, Hajime. Director of Chisso Hospital. Patient records.

Hightower, Jane M., MD. *Diagnosis Mercury: Money, Politics & Poison.* Washington/London: Island Press/Shearwater Books, 2009.

"History of Minamata." *http://minamatacity.jp/eng/history.htm.* Accessed 1/5/2009.

Jernelov, Arne. "Iraq's Secret Environmental Disasters." *www.project-syndicate.org/print_commentary/jernelov3/English.* Accessed 1/5/2009.

Justus, Sandra. "Mercury Poisoning: The Minamata Tragedy." *Environmental Biology* 11/27/98. *http://members.tripod.com/~Sandra_Justus/MercuryPoisoningReport.html.* Accessed 12/3/2008.

Kay, Jane. "Mercury in fish poses heart threat for middle-aged men, study says." *SF Chronicle* 2/8/05. SFGate.com. Accessed 3/8/2009.

———. "Bay Area New Plan to cut Mercury Release into Bay." *SF Chronicle* 5/1/04. SFGate.com. Accessed 1/5/2009.

Kim, Hyojin. "Mercury and Methylmercury in the San Francisco Bay area: Land-use impact and indicators." California Water Symposium, May 16, 2008.

"Levels of Methylmercury and Omega-3 Fatty Acids." American Heart Association. *www.americanheart.org/print_presenter.jhtml;jsessionid =TUPZLHPN4041MCQFC.* Accessed 12/3/2008.

Marsden, William. "Great lakes Basin Chemical Pollution Threatens Millions." *The Montreal Gazette* 2/14/08. *www.canada.com/ottawacitizen/news/story.html?id=9a3b7363-1935-4046-817d-f8623c2db841.* Accessed 12/03/2008.

Masazumi, Harada, *Minamata Disease.* Kumamoto, Japan: Kumamoto Nichinichi Shinbun Centre & Information Center/Iwanami Shoten Publishers, 1972.

"Mercury." *www.epa.gov/mercury/.* Accessed 1/8/2009.

"Mercury." *www.atsdr.cdc.gov/toxicprofiles/tp46-05.pdf.* Accessed 1/15/2009.

"Mercury in the Environment." Fact Sheet 146-00. 10/2002. U.S. Geological Survey. *www.usgs.gov/themes.factsheet/146-00.* Accessed 1/8/2009.

"Mercury Levels in Commercial Fish and Shellfish." U.S. Department of Health and Human Services and U.S. Environmental Protection Agency. 2/2006. *www.cfsan.fda.gov/~frf/sea-mehg.html.* Accessed 1/17/2009.

"Mercury Toxicity and How it Affects our Health." *www.mercurysafety.co.uk/hlthinfo.htm.* Accessed 12/3/2008.

"Minamata City Hall Environmental Policies." *www.minamatacity.jp/eng/cityhall_env_pol.htm.* Accessed 12/26/2008.

"Minamata Disaster." *TED Case Studies #246. www.american.edu.TED.MINAMATA.HTM.* Accessed 12/3/08.

"Minamata Disease." *www.absoluteastronomy.com/topics/Minamata_disease.* Accessed 1/8/2009.

Nobuo, Miyazawa. "Minamata Disease: A History of Japanese Government and Kumamoto Prefectural Irresponsibility." 10/15/ 2001. *http://aileenarchive.or.jp/minamata_en/aboutminamata/index.html.* Accessed 12/5/2008.

Palmer, Chad, et. al. "Who's Emitting Mercury?" *www.usatoday.com/news/mercury-emitter-map.htm.* Accessed 10/6/2008.

"Pregnancy Eating Guidelines." *http://healthytuna.com/health-nutrition/tuna-all-ages/pregnancy-eating-guidelines.* Accessed 2/02/2009.

Radsky, Oken E., JS, et al. "Maternal Fish Intake during Pregnancy, Blood Mercury Levels, and Child Cognition at Age 3 in a US Cohort." *American Journal of Epidemiology* 167 (May 15, 2008): 1171–1181.

Rellick, Gerald. "George Bush on Mercury Pollution: Don't Eat Tuna Fish." *Intervention* April 4, 2005. *www.infowars.com/articles/science/mercury_pollution_bush_dont_eat_tuna.htm*. Accessed 1/14/2009.

Sanfeliu, et al. "Neurotoxicity of Organomercurial Compounds." *Neurotoxicity Research* 5(4) (2003): 283–305.

Schertow, John. "Bush Signs into Law Obama-Murkowski-Allen Bill to Ban Dangerous Mercury Exports." 7th Space Interactive. 10/17/2008.

———. "Court Rules EPA Violated the Law by Evading Required Power Plant Mercury Reductions." Environmental Defense Fund, February 8, 2008. *http://fuel-efficient-vehicles.org/energy-news/?p=488*. Accessed 1/07/2009.

———. "Great Lakes: Chemical Soup." *Canada and the World Backgrounder* May 1, 2001.

———. "Mad as a Hatter: Canada's Mercury Pollution on Indigenous Lands." *The Dominion* August 15, 2008.

Vedantam, Shankar. "EPA Inspector Criticizes Mercury Plan." *The Washington Post* 2/4/05.

Waite, Jim. "Challenging Federal Mercury Rollbacks." SELC Cases. 12/19/08. *www.southernenvironment.org/cases/challenging_federal_mercury_rollbacks*. Accessed 1/7/2009.

11

Abdussamatov, Habibulto. "The Sun Defines the Climate." Climate Realists Online. *http://climaterealists.com/index.php?id+4254*. Accessed 12/1/2009.

Dengel, Ingrid, Dominick Aeby, and John Grace. "Á Relationship Between Galactic Cosmic Radiation and Tree Rings." *New Phytologist* 184 (November 2009): 545–551.

Idson, Craig, and S. Fred Singer. "Climate Change Reconsidered: 2009 Report on the Nongovernmental Panel on Climate Change." (NIPCC) The Heartland Institute. *www.heartland.org/publications/NIPCC%20report/PDFs/NICPP20%Final.pdf.*

"UN Climate Scientists Speak Out on Global Warming." *The Science and Public Policy Institute SPPI Reprint Series* (September 8, 2009). *http://hatch.senate.gov/public/_files/UNClimateScientistsSpeakOut.pdf.* Accessed 4/22/2010.

12

Baty, Phil. "Scientists Fail to See Eye to Eye Over Girl's X-Ray Vision." *The Higher Times Educational Supplement* 12/10/2004 *www.thes.co.uk/story?aspx?story_id=201825.* Accessed 12/22/2008.

Bauer, Eberhard. "Criticism and Controversy in Parapsychology." *European Journal of Parapsychology* 5 (1984): 141–166.

Bem, Daryl J. "Ganzfeld Phenomena." *Encyclopedia of the Paranormal.* Buffalo: Prometheus, 1996.

———. "Response to Hyman." *Psychological Bulletin* 115, no. 1 (1994): 25–27.

Bem, Daryl, John Palmer, and Richard Broughton. "Updating the Ganzfeld Database: A Victim of its Own Success?" *The Journal of Parapsychology* 65 (September 2001).

Bem, Daryl J., and Charles Honorton. "Does Psi Exist? Replicable Evidence for an Anomalous Process of Information Transfer." *Psychological Bulletin* 115:1 4–18.

Cosgrove-Mather, Bootie. "Poll: Most Believe in Psychic Phenomena." *CBS Evening News* 4/28/2002. *www.cbsnews.com/stories/2002/04/29/opinion/polls/main507515.shtml.* Accessed 6/15/2009.

"Girl With X-ray Eyes, The." *www.worldupliftment.org/super2.html.* Accessed 12/29/08.

Honorton, Charles, and Ray Hyman."The Ganzfeld PSI Experiment: A Critical Appraisal." *Journal of Parapsychology* 50 (1986).

Hyman, Ray. Comment in *Statistical Science* 6 (1991): 392.

Josephson, Brian. "Scientists' Unethical Use of Media for Propaganda Purposes." University of Cambridge. *www.tcm.phy.cam. ac.uk/~bdj10/propaganda/index.html.* Accessed 6/18/2009.

Palmer, John. "An Updated Meta-Analysis of the Post PRL ESP-Ganzfeld Experiments: The Effect of Standardness." Institute of Parapsychology. *http://parapsykologi.se/artiklar/e-ganzmetapa-b. html.* Accessed 6/13/2009.

Radin, Dean PhD. *The Conscious Universe.* San Francisco: Harper Collins, 1997.

———. "Event Related Electroencephalographic Correlations Between Isolated and Human Subjects." *The Journal of Alternative and Complementary Medicine* 10:2 (2004): 315–323.

———. "A Field Guide to Skepticism: Why the Skeptics Don't Give Up." *www.skepticalinvestigations.org/guide/.* Accessed 6/24/2009.

———. "Scientific Evidence for Psi Phenomena." *www.skepticalinvestigations.org/New/Examskeptics/Radin_guide .html.* Accessed 6/15/2009.

Rhine, J.B. "Excerpts from a Brief Introduction to Parapsychology." *Parapsychology and the Rhine Research Center.* Durham, N.C.: The Parapsychology Press, 2001.

"Russian X-ray girl Natasha Demkina still uses her gift to help common people." *Pravda* 5/29/2008. *http://english.pravda.ru/science/ systeries/105380-Pravda.Ru-natasha_demkina-0.* Accessed 12/23/2008.

"Russian x-ray girl thrills Japanese scientists with her remarkable gift." *Pravda* 2/4/05. *http://english.pravda.ru/science/health/8097-demkina-0.* Accessed 12/23/2008.

Sample, Ian. "Visionary or Fortune Teller? Why Scientists Find Diagnoses of 'X-Ray' Girl Hard to Stomach." *The Guardian* September 25, 2004 *www.guardian.co.uk/uk/2004/sep/25/russia.health.* Accessed 12/22/2008.

Skolnick, Andrew A. "Natasha Demkina: The Girl with Very Normal Eyes." *Skeptical Inquirer www.csicop.org/si/2005-05/demkina.html.* Accessed 12/22/2008.

Utts, Jessica, PhD. "Replication and Meta-Analysis in Parapsychology." *Statistical Science* 6, no. 4, (1991): 363–403. *www.stat.ucdavis. edu/~utts/91rmp.html.* Accessed 6/24/2009.

———. "Response to Ray Hyman's Report." 9/11/1995 "Evaluation of Program on Anomalous Mental Phenomena." 9/15/1995 *http://anson.ucdavis.edu/~utts/response.html.* Accessed 12/25/2008.

Utts, Jessica, and Brian D. Josephson. "The Paranormal: The Evidence and its Implications for Consciousness." 4/5/1996. *www.tcm.phy. cam.as.uk/~bdj10/psi/tucson.html.* Accessed 12/23/2008.

Zammit, Victor. "Natasha Demkina can sue the 'Ambush Experimenters' for some $2 to $5 Million in Damages. Why?" *http://victorzammit.com/articles/natashacansue.html.* Accessed 12/27/2008.

———. "Victor Zammit on Medical Intuitive Natasha Demkina: A most disturbing and professionally offensive psychic experiment." *www.victorzammit.com/articles/natasha.html.* Accessed 12/27/2008.

13

AIAA UFO Subcomittee. *UFO: A Scientific Appraisal of the Problem.* Astronautics and Aeronautics 8:11 (1970): 49.

Asimov, Isaac. "The Rocketing Dutchman." *Fantasy and Science Fiction* (February 1975): 132.

Bolender, Brigadier General Carroll H. Memo. *Unidentified Flying Objects* October 20, 1969.

Condon, Edward U. *Scientific Study of Unidentified Flying Objects.* New York: Bantam Press, 1969.

Feschino, Frank Jr. *Shoot Them Down. www.stantonfriedman.com,* 2007.

Friedman, Stanton T. *Flying Saucers and Science.* Franklin Lakes, N.J.: New Page Books, 2008.

———. *TOP SECRET/MAJIC 2nd Edition.* New York: Marlowe and Co., 2005.

Fuller, John G. "The Flying Saucer Fiasco." *Look Magazine* May 14, 1968.

Hynek, J. Allen. *The UFO Experience: A Scientific Inquiry.* Chicago: Henry Regnery, 1972.

Jacek, Martin. *UFOBC Special Report: Giant UFO in the Yukon Territory* June 2000.

Luce, John S. *Controlled Fusion Propulsion, Proceedings of the Third Symposium on Advanced Propulsion Concepts* Vol. 1. Amesterdam: Gordon and Breach Science Publishers, 1963.

McCullough, David. *Truman.* New York: Simon and Schuster, 1992.

McDonald, James E. "RB-47 Case Report for the AIAA." *Astronautics and Aeronautics* July 1971.

———. *Statement on UFOs to United States Congress* July 29, 1968.

Project Blue Book Special Report No. 14. Prepared by Battelle Memorial Institute for United States Air Force Project Blue Book, 1955.

Sagan, Carl. *The Cosmic Connection: An Extraterrestrial Perspective.* New York: Anchor Press, 1973.

———. *The Demon-Haunted World: Science as a Candle in the Dark.* New York: Random House, 1996.

Sagan, Carl, and Ann Druyan. *The Varieties of Scientific Experience: A Personal View of the Search for God.* New York: Penguin Press, 2006.

Sagan, Carl, and Thornton Page. *UFO's: A Scientific Debate.* Ithaca, N.Y.: Cornell University Press, 1972.

Symposium on Unidentified Flying Objects. House Committee on Science and Astronautics. July 29, 1968.

14

Appelle, Stuart. "The Abduction Experience: A Critical Evaluation of Theory and Evidence." *Journal of UFO Studies* n.s. 6 (1995/96): 29–78.

Bartholomew, Robert E., Keith Basterfield, and G.S. Howard. "UFO Abductees and Contactees: Psychopathology or Fantasy Proneness?" *Professional Psychology: Research and Practice* 22, no. 3. (1991): 215–22.

Beidas, Rinad. "Individual Differences in the Formation of False Memories: Is Suggestibility a Predictive Factor?" *Colgate University Journal of the Sciences* (no date given) 77–91.

"Borderline Personality Disorder." *www.palace.net/~llama/psych/bpd.html.* Accessed 7/5/2009.

Blackmore, Susan. "Abduction by Aliens or Sleep Paralysis?" *Skeptical Inquirer Magazine* May/June 1998. *www.ufoevidence.org/documents/doc817.htm.* Accessed 7/2/2009.

Boylan, Richard, PhD. "The Differential Diagnosis of Close Extraterrestrial Encounter Syndrome." *www.drboylan.com/diffdxce4a.html.* Accessed 7/2/2009.

Bryan, C.D.B. *Close Encounters of the Fourth Kind: Alien Abduction, UFOs and the Conference at MIT.* New York: Alfred A. Knof, 1995.

Clancy, Susan A. *Abducted: How People Come to Believe They Were Abducted by Aliens.* Cambridge: Harvard University Press, 2005.

———. et al. "Memory Distortion in People Reporting Abduction by Aliens." *Journal of Abnormal Psychology* 3, no. 3 (2002): 455–61.

———. et al. "Psychophysical Responding During Script-Driven Imagery in People Reporting Abduction by Space Aliens." *American Psychological Society* 15, no. 7 (2004): 493–97.

Condon, Edward U. *Scientific Study of Unidentified Flying Objects.* New York: Bantam Books, 1969.

Donderi, Don C, PhD. "The Scientific Context of the UFO/Abduction Phenomenon." December 28, 1996. *www.ufologie.net/htm/donderi.htm.* 7/2/2009.

———. "Evidence Weakens Psychological Theories of the Alien Abduction Experience." McGill University Department of Psychology. Montreal, Canada. 2008 presentation.

Friedman, Stanton. "The Pseudoscience of Antiufology." *Pursuit* (1983): 17–20.

Friedman, Stanton, and Kathleen Marden. *Captured! The Betty and Barney Hill UFO Experience.* Franklin Lakes, N.J.: New Page Books, 2007.

Fuller, John G. *The Interrupted Journey*. New York: Dial Press, 1996.

Hopkins, Budd. *Art, Life and UFOs: A Memoir*. New York: Anomalist, 2009.

———. "The Faith-Based Science of Susan Clancy." *www.intrudersfoundation.org/faith_based.html*. Accessed 7/9/2009.

———. *Intruders*. New York: Random House, 1987.

———. *Witnessed*. New York: Pocket Books, 1996.

Hopkins, Budd, and Carol Rainey. *Sight Unseen*. New York: Atria, 2003.

Howe, Linda Moulton. *Glimpses of Other Realities: Volume I : Facts and Eyewitnesses*. Huntingdon Valley: LMH, 1993.

———. *Glimpses of Other Realities. High Strangeness Volume II*. New Orleans, La: Paper Chase Press, 1998.

Jacobs, David Michael, PhD. "A Review of *Abducted: How People Come to Believe they were Kidnapped by Aliens* by Susan Clancy." *The Journal of Scientific Exploration* 20, no. 2 (Summer 2006).

———. *The Threat: The Secret Alien Agenda*. London: Simon & Schuster, 1998.

———, ed. *UFOs and Abductions: Challenging the Borders of Knowledge*. Lawrence, Kansas: Kansas University Press, 2000.

Klass, Phillip J. *UFO Abductions: A Dangerous Game*. Buffalo, New York: Prometheus, 1989.

Kottmeyer, Martin. "Abductions: The Boundary Deficit Hypothesis 1988." *www.think-aboutit.com/abductions/TheBounraryDeficitHypothesis*.   Accessed 7/6/2009.

Lagrandeur, Tamara, Don C. Donderi, Stuart Appelle, and Budd Hopkins. "Self-Reported Alien Abductees Remember Consistent Sets of Symbols." PowerPoint presentation. Chicago. May 23, 2008.

Leir, Roger K. *Casebook: Alien Implants*. New York: Dell, 2000.

———. Lecture at the 2009 International UFO Congress and Film Festival.

Lindemann, Debra L. "Surgeon Tells First Results of Implant Analysis." 1996. *www.v-j-enterprises.com/iscni24.html*. Accessed 7/7/2009.

Low, Robert. Memo to James Archer, the dean of the University of Colorado, dated 8/9/1966.

Mack, John E. *Abduction: Human Encounters with Aliens.* New York: Scribner's, 1994.

———. Caroline McLeod, and Barbara Corbisier. "A More Parsimonious Explanation for UFO Abduction." *Psychological Inquiry* 7, no. 2 (1996): 156–68.

Marden, Kathleen. "Another Look at the Harvard False Memory Study." March 23, 2006.

———. "Barney and Betty Hill: Dream Transference or Alien Abduction?" *UFO Journal* (April 2009).

National Academy of Sciences, "Review of the University of Colorado Report on Unidentified Flying Objects by a Panel of the National Academy of Sciences" (1969). *www.project1947.com/shg/articles/nascu.html.* Accessed 7/2/2009.

Parnell, June O. and R. Leo Sprinkle. "Personality Characteristics of Persons Who Claim UFO Experiences." *Journal of UFO Studies* 2 (1990): 45–58

Powers, Susan Marie. "Fantasy Proneness, Amnesia and the UFO Abduction Phenomenon." *Dissociation* IV, no 1 (March 1991).

Randle, Kevin D., Russ Estes, and William P. Cone, PhD. *The Abduction Enigma.* New York: Tom Doherty Associates, 1999.

Ring, K. and C. Rosing. "The Omega Project: A psychological survey of persons reporting abductions and other UFO encounters." *Journal of UFO Studies* 2 (1990): 59–98.

Rodeghier, Mark. "Abducted by her Beliefs." *IUR* 30:2 (January 2006).

Rodeghier, M., J. Goodpastor, and S. Blatterbaur. "Psychosocial Characteristics of Abductees: Results From the CUFOS Abduction Project." *Journal of UFO Studies* 3 (1991): 59–90.

Sagan, Carl. "Are They Coming for Us?" *Parade* March 7, 1993.

Saunders, David R., and R. Roger Harkins. *UFOs? Yes!* New York: Signet, 1968.

"Schizotypical Personality Disorder." *http:// www. Mentalhealth.com/ dis1/p21-pe03.html*. Accessed 3/10/2006.

Sims, Derrel. *Alien Hunter: The Evidence in Light*. Houston: D.W. Sims, 2006.

———. *Alien Hunter: Implants: Evidence and Truth about Alien Implants*. *www.alienhunter.org/store/implants-as-evidence.html*.

Spanos, Nicholas P., et al. "The Social Reconstruction of Memories." *www.psywww.com/asc/hyp/memories.html*. Accessed 7/9/2009.

Spanos, N., P. Cross, K. Dickson, and S. DuBreuil. "Close Encounters: An Examination of UFO Experiences." *Journal of Abnormal Psychology* 102 (1993): 624–32

Webb, Walter. "A Dramatic UFO Encounter in the White Mountains, NH." Confidential NICAP Report. October 26, 1961.

Whittlesea, Bruce, Michael E.J. Masson, and Andrea D. Hughs. "False Memory Following Rapidly Presented Lists: the Element of Surprise." *Psychological Research* 69 (2005): 420–430.

Wilson, Sheryl C. and Theodore X. Barber. "The Fantasy Prone Personality: Implications for Understanding Imagery, Hypnosis, and Paraphysiological Phenomena." *Imagery, Current Theory, Research and Applications*. New York: John Wiley & Sons, 1983. 340–390.

Index

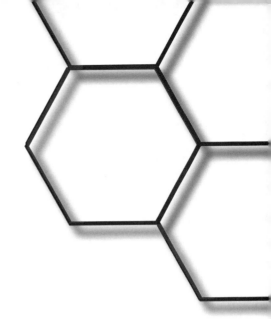

About the Authors

Stanton T. Friedman

Nuclear physicist and lecturer Stanton T. Friedman received his BSc and MSc degrees in physics from the University of Chicago in 1955 and 1956. He was employed for 14 years as a nuclear physicist by such companies as GE, GM, Westinghouse, TRW Systems, Aerojet General Nucleonics, and McDonnell-Douglas, working on such highly advanced, classified, eventually cancelled programs as nuclear aircraft, fission and fusion rockets, and various compact nuclear powerplants for space and terrestrial applications.

He became interested in UFOs in 1958, and has lectured about them at more than 600 colleges and 100 professional groups in all 50 states, nine Canadian provinces, and 16 other countries, in addition to various nuclear consulting efforts. He has published more than 90 UFO papers and has appeared on hundreds of radio and television programs, including on *Larry King Live* in 2007 and twice in 2008, and many documentaries. He is the original civilian investigator of the Roswell Incident and coauthored *Crash at Corona: The Definitive Study of the*

Roswell Incident. TOP SECRET/MAJIC, his controversial book about the Majestic-12 group established in 1947 to deal with alien technology, was published in 1996 and went through six printings. An expanded new edition was published in 2005. Stan was presented with a Lifetime UFO Achievement Award in Leeds, England, in 2002, by *UFO Magazine* of the UK. In 2007, he coauthored, with Kathleen Marden (Betty Hill's niece), *Captured! The Betty and Barney Hill UFO Experience.* The City of Fredericton, New Brunswick, declared August 27, 2007, Stanton Friedman Day. His new book, *Flying Saucers and Science,* was published in June 2008 and is in its 4th printing.

He has provided written testimony to congressional hearings, appeared twice at the UN, and been a pioneer in many aspects of ufology, including Roswell, Majestic-12, the Betty Hill–Marjorie Fish star map work, analysis of the Delphos, Kansas, physical trace case, crashed saucers, flying saucer technology, and the challenges to S.E.T.I. (Silly Effort To Investigate) cultists. He as spoken at more Mutual UFO Network Symposia than anyone else.

Stanton T. Friedman is a dual citizen of the United States and Canada, and can be reached at fsphys@rogers.com or *www.stantonfriedman.com.*

Kathleen Marden

Kathleen Marden was trained as a social scientist and educator and holds a BA degree from the University of New Hampshire. She participated in graduate studies in education at The University of Cincinnati and UNH. She is also a certified hypnotherapist. During her 15 years as an educator, she innovated, designed, and implemented model educational programs. She also held a supervisory position, coordinating, training, and evaluating education staff. Additionally, she taught adult education classes on UFO and abduction history. She is the director of field investigator training (emeritus) for the Mutual UFO Network, the largest international UFO organization. In 2003, MUFON publicly recognized Kathy for her outstanding contribution to advancing the scientific study of the UFO phenomenon and demonstrating positive leadership.

For the past 20 years, she has engaged in UFO research and investigation and written numerous articles pertaining to UFO abduction phenomena. Kathy was a primary witness to the evidence of her aunt and uncle Betty and Barney Hill's 1961 UFO encounter and the aftermath. After Betty's death in 2004, Kathy compiled two permanent archival collections for the Milne Special Collection and Archives Department at the UNH Library. The Hills' civil rights collection is composed of documents, letters, photographs and newspaper articles pertaining to Betty's and Barney's social and political activities. The UFO collection contains all of the correspondence, articles, and other material from Betty's extensive files, including new material, Betty's dress, and the forensic paintings of her captors by artist David Baker.

In 1996, Betty released to Kathy the audio tapes of the hypnosis sessions she and Barney participated in with renowned psychiatrist Dr. Benjamin Simon. With Betty's approval, Kathy transcribed the tapes and conducted a comparative analysis of the Hills' individual testimony versus Betty's dream account of alien abduction. Her book *Captured! The Betty and Barney Hill UFO Experience* (coauthored with nuclear physicist/scientific ufologist Stanton Friedman) was published in 2007. Kathy has appeared on numerous television and radio programs in the United States, Canada, and Europe. Additionally, she has lectured throughout the United States.

Kathy divides her time between Clermont, Florida, and Newburyport, Massachusetts. She can be reached at Kmarden@aol.com or at her Website *www.kathleen-marden.com.*